GREEN RACING

A Case Study of South Korea

The Energy Story of a Manufacturing Powerhouse

PROF. KIM CHANGSEOP

INDIA • SINGAPORE • MALAYSIA

ISBN 979-8-89026-859-4

Table of Contents

References 5

Prologue 7

Part 1: Energy has been with us for a Long Time 13

Chapter 1: What is Energy? A Very Long Story 17

Chapter 2: There are Many Factual Stories About Each Energy Source 21

Chapter 3: Korea's Energy Story 31

Chapter 4: Let's Learn About the Climate Agreement 36

Part 2: Maintain the Status of a Manufacturing Powerhouse as Carbon Neutral 48

Chapter 1: Korea is Now Clearly an Advanced Country 53

Chapter 2: A New Huge Crises is Around the Corner 62

Chapter 3: The Public and Private Sectors Must Work Together in Certain Manufacturing Industries 69

Chapter 4: Why is Carbon Neutrality a Core Policy of Manufacturing Powerhouses? 74

Part 3: Carbon Neutrality Breaks Through Innovation in Energy Technology 79

Chapter 1: How to Respond to the Climate Agreement 83

Chapter 2: Climate Response and Breakthrough Technologies to Protect Manufacturing Powers 92

Chapter 3: Greening of Manufacturing Capacity and Energy Policies have Already Grown Together 102

Chapter 4: Green Racing's Energy Policy is a New Industrial Policy. 107

Part 4: Three Enemies Preventing Green Racing.............112

Chapter 1: Populists Who Insist on Cheap Energy are Enemies! 114

Chapter 2: 'Monocolorists' Obsessed with Specific Energy Sources are Also Enemies!................................. 132

Chapter 3: The Confusion of too Many Old Energy Plans is Also an Enemy.. 141

Chapter 4: There are Groups that Must be Persuaded to Join 148

Part 5: A Reflection on Green Racing and More............. 152

Chapter 1: The Energy Policy Must be Improved in Four Basic Directions 154

Chapter 2: Integrates Complex Energy Market Rules in an Easy-To-Understand Manner 163

Chapter 3: Green Racing Also Requires Strategic Thinking 172

Finishing Here 178

References

The Colours of Hydrogen

First Five-Year Economic Development Plan

U.S. IRA Act

RE100

CBAM

Plaza Agreement

Data Pictures (Energy Density)

Evidence that Korea, China, and Japan were the greatest beneficiaries

Carbon Neutrality

NDC

Climate Agreement, Paris Agreement

TIPPING POINT THEORY

NDC Content Submitted by Korea

2050 Core Technologies Proposed by Net Zero Forum

2 CFI Pictures

Energy Costs for Current Consumers

Consumers in Korea are Procuring Energy Costs in Various Forms – Electricity Consumers, Petroleum Consumers, District Heating Consumers, LPG Consumers, etc. As a Result, Various Financial Structures Have Occurred. For Example:

- Fee and Tax Structure
- Foreign Exchange Fees Paid by Korean Consumers
- Oil and Gas Imports, etc.

Comparing Energy Costs by Country

Externality Theory and Data

GRID PARATY

The Shift in the Price Balance Point That Occurs Every Time Fuel is Switched

Legal Plans Related to Energy and Climate

Markets CBP, ETS, RPS, HPS, etc.

Prologue

We have become a developed country after 60 years of hard work. We have realized the dream of industrialization and simultaneous democratization that began in 1962. It has been said that this success is close to a miracle in the history of humankind. With the advent of globalization and digitalization, I woke up one day and found that I was in a developed country. We are fulfilling our dreams not only in automobiles, semiconductors, and shipbuilding, but also in cultural and artistic fields, what with the success of BTS, K-drama, and K-pop. Furthermore, K9 self-propelled artillery, K2 tanks, and FA50 fighters are sold all over the world, including Europe. I'm yet to realize that these achievements are amazing. However, international external conditions that were very favourable to Korea, such as globalization and oil price stability, are changing. Our internal patriotism, which was our competitive advantage, and culture, are rapidly changing due to the saturation of infrastructure-building such as power plants and power grids, and the deepening of various conflicts. In other words, our way of achieving success is under threat. From this perspective, I'm most curious about whether our success and wealth will be maintained in the future. What kind of crisis awaits us? Can we overcome it? What level of empathy is needed?

Meanwhile, what is the source of our wealth? It is our manufacturing capacity. The agricultural and financial sectors also have great potential to grow in the future, but our main industry is still the manufacturing sector. This was the route taken by the Republic of Korea 60 years ago. We began our path to industrialization on January 13, 1962, when we announced the First Five-Year Economic Development Plan. Although there were many debates and conflicts raging, the policy goals of the heavy chemical industry were achieved through a period of development dictatorship. This is the truth. Based on this, high-value-added industrialization such as the manufacture of semiconductors was achieved. There are amazing stories in the sports and cultural sectors that are even more exciting. Considering that Korea was a miserable agricultural country where everything was destroyed by war in 1962, the past 60 years have proved to be the age of heroes.

On the other hand, there's something we've overlooked. An efficient energy policy lies in the success of the manufacturing industry. Both began together in 1962. In the spring of 1962, the Republic of Korea announced its first long-term power development plan and proactively began measures to procure the energy required for the five-year plan. The story of Bukuk Kang Byeong of the Republic of Korea was realized by harmonizing the mythology of the heavy chemical industry with a sincere energy policy. As a result, we were able to build the best high-density energy society in human history even though we were a resource-poor country. Also, while maximizing manufacturing capacity, Korea paid close attention to nature and the environment. When I watched Korean movies from the 50s and 60s, most of the mountains in the background were the Mindong Mountains. However, we planted one tree at a time to create the current greenery. When I was a child, all students and office workers were mobilized to plant seedlings on Arbor Day, and when the time came, they would go out to catch insects. In addition, they protected nature by maintaining policies that propagated the Green Belt Movement for a considerable period. We also have experience in greening. Thus, it is clear that Korea is an amazing and dynamic country focused on innovation.

However, conditions have changed drastically, and a new crisis is emerging. The disappearance of free trade and the restructuring of the supply chain due to the conflict between the US and China is unsettling for us as manufacturers of import/export countries. Furthermore, various regulations have been put in place in response to the climate change crisis. We need to fundamentally revise the energy system we have been used to. This involves huge costs, and it's quite possible that the stable supply and demand of energy will be threatened. This is particularly a serious challenge for those of us who are poor in energy resources and have an energy-intensive industry. This crisis is not local; it is global. It will be perpetuated by triggering large and small disputes. As shown in the movie *Tomorrow*, it could turn into a battle between countries for land to meet their future needs.

It is necessary that we prepare ourselves to overcome this obvious crisis in front of us. Only in this way can we promote our wealth and safety in the long run. Right now, we are unable to respond appropriately. This is because each of us is already struggling daily to overcome the present crisis in our difficult lives. We continue to stick to the methods of the past. Meanwhile, the crisis is only becoming worse with time. Responding to this crisis on an individual level is clearly limited. The whole community must work together. However, I'm extremely concerned when I see the current political climate engulfed by strife and conflict. It makes me reconsider the enforced patriotism of the past era of authoritarian development, where individual lives and communal existence were seen as a collective destiny. However, the new leadership in developed countries is even more disappointing. What is clear is that individuals cannot overcome this crisis. Nobody has wings when falling.

There's a point that needs to be clarified here. Climate change will not lead to the annihilation of humanity. Any country or class can see virtue. At the end of the movie *Tomorrow*, we see the US President begging the Mexican President to open borders on the condition that all debts are forgiven. Of course, I'm guessing that the second Texas War may occur in the sequel. After all, climate change is a challenge, but it must be made clear that it does not mean annihilation. It has been implied that homo sapiens, who lived in Africa long ago, spread all over the world due to climate change and became the dominant species. We are entering an era of confusion where someone becomes the loser, and someone else becomes the winner. Where is our country heading to?

Of course, most citizens are also having serious discussions on climate change, looking at the situation from various perspectives. Some of them are concerned about polar bears or the global environment. I have never worried about polar bears. Also, Gaia would be puzzled to hear some people's concerns about her. From Gaia's point of view, humanity, which runs on the capitalism of mass production and consumption, is probably like an ant that bothers her by mining oil and making holes in her body. Therefore, there is a great possibility that humanity will suffer due to climate change and drinking bone soup will be cool. As abnormal temperatures have already occurred across the world, climate response is in full swing. Climate response is now being put into practice. We need a realistic and new understanding of whose interests will be prioritized. It's not a situation for others to worry about. Of course, I believe that the interests of the Republic of Korea should be considered. National interests come first. In the end, it will benefit each of our citizens, and will also be right for Gaia.

In fact, countries around the world have already begun fierce competition over national interests other than humans and polar bears. Climate-related global regulations such as RE100 and CBAM are also being further tightened. However, the situation began to intensify when the US enacted the IRA law. This is an effort by the United States to respond to the climate crisis while simultaneously moving climate-related factories from around the world to the United States. Europe opposed it by claiming that it was a violation of the WTO spirit, but they are also responding by enacting the same laws. The competition is intensifying. A war has begun over who will equip themselves with a green transformation more quickly, attracting manufacturing plants and jobs. This is called 'green racing.' Although the term 'green racing' isn't commonly used, I think it's an apt expression in the current situation. However, I would like to call it 'green racing,' rather than green racing, in the sense that it is currently going on fiercely.

I will explore these stories in more detail from Part 1 through Part 5. Let's look at what we've achieved and what kind of crisis we're facing. I will talk about our most serious internal enemies in responding to this crisis. I will propose very basic countermeasures to respond to the crisis. I will examine how the energy system should change to turn this crisis into an opportunity for the sustainable growth of the Korean economy. Discussions on how energy policies should be integrated with industrial policies are also critical.

Part 1: Energy has been with us for a Long Time

In ancient times, when Prometheus gifted humanity with fire or energy, the development of civilization began. The brain grew with the advent of paper, which brought about the industrial revolution during the Bronze and Iron Ages when capitalism finally had its beginnings. It's safe to say that it all started with energy. As humans used this energy, the composition of the Earth's atmosphere changed. Of course, the composition of the Earth's atmosphere has been constantly changing. However, this is the first time in the Earth's 4.5-billion-year history that this composition has been rapidly changing, apart from the impact of a rare asteroid.

Energy has improved human life over the years. The sources of energy are diverse. Most are useful resources through collaboration between the Sun and Earth. We have created today's incredible civilization by utilizing various energies such as coal, oil, gas, solar energy, and wind power. Furthermore, the Republic of Korea, which barely had any resources, built the world's highest-density energy society. However, I wonder what Grandpa Dan-gun will say when he sees it. It's also a precarious achievement that can be easily ruined as soon as an oil transportation route is blocked. We must acknowledge how extraordinary our achievements have been, but we must also be aware of how precarious they are. It is true that we lack public understanding or consensus on energy.

Japan, which is in a situation like ours, should prioritize energy security when formulating its policies. It's worth keeping in mind that the Pacific War began with the US oil embargo. The grid energies we depend on (electricity, gas, heat, etc.) can stop at any time. Think of a situation where a power outage occurs. Citizens may experience some inconvenience in the early stages, but only after a few days will they realize that their lives have been destroyed, and they must fight to survive. The collapse of the modern Tower of Babel will have far more frightening consequences compared to the pain of the linguistic confusion at the Tower of Babel in the Bible. Energy does not simply cause human suffering due to fine dust or climate change; it is a ticking bomb that has the power to destroy people's daily lives right now. Therefore, it is possible that the Gods punished Prometheus for his actions.

We need to understand what energy is. Also, let's look at the history of energy in Korea. Only then will the future national challenges we must tackle be more clearly understood. Since our energy policies have always been better than any other country so far, our consumers are aware that energy is a natural waste like air. However, this is not the case. Even though Korea has virtually no energy resources, it is the world's largest energy consumer. The strength of one era becomes the weakness of the next. That's where we are now.

The hardships may be particularly difficult for consumers who cannot endure a power outage of several hours. It is difficult to take for granted the best energy service at the lowest cost. The 915 Great Power Outage during the famous government of Lee Myung-bak was not actually a power outage, but a circular power outage without warning, and was a problem that could not be discussed in other countries. However, as a result of this minor incident, officials from the Central Government and related agencies were questioned at an unprecedented level. As such, energy has become a natural right. So, it seems reasonable that politicians are extremely averse to increases in energy bills. Now this extravagant period is over. However, no one talks about such inconveniences to consumers. In an age of populist judgement, this truth seems to be accepted as an act of self-harm that eats away at votes. However, we now live in an age where it is difficult to build power plants, transmission towers, substations, and garbage dumps. I emphasize the need for nuclear power plants, but I want to avoid discussions about the final disposal of settled nuclear waste. Eventually, our quality energy services will stop due to a lack of infrastructure. The normal operations of the power plant that have already been built on the East Coast are in jeopardy. Furthermore, due to fine dust and climate responses, these overall changes in the energy system are inevitable. The good times have come to an end.

I would like to share a general story about energy and my memories of energy in Korea. We need to know how well we've built energy. It's a source of pride. Also, it is necessary to understand how they are linked to climate issues. This is because we can act only if we understand the situation. It is essential to have this awareness to respond to the climate in the future. It impacts our survival, too.

Chapter 1

What is Energy? A Very Long Story

As everyone knows, a long time ago, Prometheus secretly gave fire to humans, and subsequently, humans became powerful. It is said that homo sapiens and Neanderthals used fire to grill meat, which improved their digestion and consequently developed their brains. Homo sapiens then used this fire to make bronze and iron tools. At last, humanity had unequivocally become the absolute powerhouse of the Earth. We call this fire energy.

Humans began using energy differently in England in the 1800s. A more concentrated energy mass called coal was put into a barrel called a steam engine and it was used to ignite a fire. Now, instead of using firewood found on the surface of the earth, humans are using fuel that has been highly concentrated underground for hundreds of millions of years. Thus, humanity was instantaneously able to utilize vast amounts of energy to produce huge quantities of goods (clothing). In sociology, this is called 'productivity.' Farmers consumed the clothes instantaneously and at the same time got jobs in factories and began to produce more clothes. However, when there were no more places to sell clothes, the capitalists began to forcibly sell them to foreign countries. This is what brought about the industrial revolution, capitalism, and imperialism. It is all due to technology that utilizes highly concentrated energy.

Energy is defined in the dictionary as "the ability to do work." We buy this energy in the form of gasoline and diesel at gas stations, buy electric energy from KEPCO to operate refrigerators, and batteries from E-Mart. We also buy ramen to fuel our bodies. It's hard for us to last a day without energy. What's different is that we must buy all our energy from the market, whether it's ramen, gasoline, electricity or batteries.

These energy resources basically come from nature. Coal and petroleum are also obtained from nature. Firewood, which is the oldest energy source, is also from nature. Such fuel sources are the result of a collaboration between the Sun God and Gaia. However, the only difference lies in how long Gaia has been embraced and how much energy concentration per unit volume has been demonstrated as a result. Wind power comes from the wind, tidal power, from the waves, and sunlight, from the sun. Of course, all these sources are pure organic, raw energy that isn't concentrated. However, there are energies that nature has not created. One such energy is nuclear energy. Some religious communities have a disrespectful view of something that goes against nature. However, it is also necessary to remember that most of the uranium, which is the raw material, has also been provided by Gaia. However, the waste generated is a huge pollutant and has been dangerous for a long time. Of course, it's true that coal and oil are also polluting the Earth and causing the sixth mass extinction. There is no relaxing energy at all.

Hydrogen energy, which has recently been gaining popularity around the world, is being discussed as an inevitable option. I believe that Korea should focus on hydrogen energy. In a nutshell, electric cars are more powerful than hydrogen cars. Hydrogen has a high energy density and is friendly to our heavy chemical industry. Even in defence, field mobility must inevitably be based on hydrogen energy. On the other hand, fusion energy, which is another option, would be a breakthrough technology if commercially possible. It has rapidly evolved in recent years, so it's worth looking into. Thus, energy is entering a new phase, being recreated by human technology. I wonder what Prometheus, who has been devouring eagles for a long time, would say about all this. I don't think I've thought about it until now. My guess is that Prometheus is now devising ways to take backfire from humans. Or the Gods may be blaming themselves for creating humans.

Of course, I believe that humanity will overcome the crises stemming from energy depletion and climate change. The homo genus species had overcome that cold ice age without any special technology or government organization. Furthermore, homo sapiens beat the muscular Neanderthals too. It is a species that is resistant to crises. We have always been dealing with some crises or the other. But can we feel the climate crisis that scientists claim will befall humanity decades from now? We have no choice but to focus on the troublesome 'everyday crises' that will come tomorrow and the day after. Consequently, climate response has continued to fail over the past 30 years. Perhaps we were able to overcome that cold and dangerous ice age by focusing at a high level on the crisis that existed in front of us. Homo sapiens must have overcome the Neanderthals the same way. The ability to focus on the immediate crisis is still alive in our DNA.

Chapter 2

There are Many Factual Stories About Each Energy Source

To briefly talk about coal, hundreds of millions of years ago, when the earth was warmed by greenhouse gases, there was a time when plants on Earth experienced the joy of optimal photosynthesis and the whole earth was filled with giant trees. Like in the Carboniferous Period? Subsequently, these giant trees were buried in the ground, and Gaia took good care of them, slowly turning them into coal. It became a completely highly concentrated energy mass. Such energy is distributed relatively equally everywhere on Earth.

Bulbs of coal have been used by humans as fuel for a long time. Then, Thomas Newcomen developed steam engines in England in 1705 and James Watt made improvements to it in 1769, sparking the Industrial Revolution. It was the rediscovery of coal. As a result, humanity ushered in an era of mass production. Then, the famous London Fog occurred, causing a terrible environmental disaster. It was a fine dust problem, but London was so dangerous at the time that it was difficult to gauge its toxicity. It's amazing that humanity was already aware of the disharmony between energy and the environment from the beginning. Essentially, coal emits a huge amount of carbon dioxide. Even now, the greatest contribution to fine dust and climate change is coal or coal power generation. Currently, it is the energy most hated by environmental groups who consider it to be the main enemy of climate response.

However, from the standpoint of coal, which has been resting comfortably in Gaia's arms for hundreds of millions of years, it is clearly a very disappointing situation. NASA has conducted research to artificially release greenhouse gases to further maintain the current interglacial period to defer the ice age that is expected to arrive soon. Research has shown that this may happen if all the coal in the world were burned. Our insight and wisdom are nothing more than a matter of magnitude.

Now, let's talk about oil. Compared to coal, the raw material for this oil is not clear. Some say it is the deposits on the pore layer, and others say it is the transformation of a mass of plankton corpses. However, we know that it was caused by Gaia holding the deposited clumps of organic matter over a long period of time. In this respect, they are like coal's brothers, collectively known as fossil fuels. Petroleum is described as bitumen in the Bible. This has also been used for a long time. When it replaced coal in steam engines, it played an important role. There is also a theory that the glory of the British Empire began in earnest when Churchill, the Minister of the Navy, changed a coal warship into a heavy ship. Petroleum has an advantage over coal.

Petroleum is also known as Devil's Tears. I wonder why? Unlike coal, petroleum is abundant in specific countries, and conflicts and wars have become frequent due to this. As a result, the discipline of geopolitics has developed further. Petroleum is important, not only as an energy source, but also as a raw material. Most of the clothes we wear are made of petroleum. Most clothing, other than silk, leather, or cotton, is made of petroleum. It's clear that if the magician had gotten rid of the oil suit right now, most of us would be naked. Plastic, vinyl, etc., are all petroleum products. A Saudi oil minister once said, "The Stone Age did not disappear because of a shortage of stones," and the British Petroleum Company was referred to as Beyond Petroleum. This generation is clearly an age of the civilization of petroleum, regardless of what people say. Just 10 years ago, the oil peak theory was the hottest thing, but now, no one is concerned that we will run out of oil resources. We will not be able to escape petroleum civilization for a long time to come. However, strangely enough, climate crisis theorists strongly criticize coal and talk little about petroleum.

Fossil fuel-based steam engines began in the United Kingdom, but electric energy began in the United States. After developing incandescent bulbs in 1879, Edison began selling commercial electricity on Wall Street for the first time in 1882, using a combination of incandescent bulbs and generators, and the era of electricity began. Korea also equipped Gyeongbokgung Palace with power generation facilities and unveiled incandescent bulbs in 1887, the first to do so in Asia. The electricity supply that began in a local neighbourhood, explosively generated demand through interconnection, leading to the construction of a full-scale transmission and distribution network. Korea furthered development by securing mass production capacity for more precise control using electric energy. Electric energy made it possible to have light at night, and washing machines drastically reduced the burden of housework for women, which was the driving force behind the rise in passports. With the advent of semiconductors and computers, humanity began the fourth industrial revolution. Electric energy not only strengthened humans but also made them smart.

An energy like electric energy but with completely different properties has emerged; this is nuclear energy. This nuclear power was developed with the atomic bomb in mind, which disgusted religious people. Nuclear power generation utilizes this heat by slowly adjusting the chain reaction of an atomic bomb. At one point, it also offered the optimism of an inexhaustible energy supply to humanity. However, in addition to risks such as the Fukushima explosion, the issue of permanent disposal of spent nuclear fuel remains a major problem for humanity and the Republic of Korea. There's a story we overlook. The first energy policy after the establishment of the Republic of Korea was nuclear energy. The Korea-US Nuclear Energy Agreement was promoted by President Lee Seung Man on February 3, 1956 (officially known as the Agreement for Cooperation between the Government of the Republic of Korea and the Government of the United States of America on the Non-Military Use of Nuclear Energy). The history of our nuclear power plants is just as old.

The energy that has recently received the most attention is renewable energy, that is, solar power and wind energy. Like its older brother, fossil fuels, the sun is the driving force, but Gaia hasn't embraced it long enough, so it's natural, unconcentrated energy. It is the next-generation energy with the highest international investment and the fastest growth rate over the past ten years. However, from the perspective of a country like South

Korea, which heavily relies on high-density energy sources, it can be challenging to transition to low-caloric organic energy. It's akin to expecting someone like Kang Ho-dong, a celebrity known for his large appetite, to survive on just grass.

Hydrogen energy is the most widely introduced energy in recent times. Strictly speaking, just as electric energy is a secondary conversion of existing energy, hydrogen energy is also converted energy. Furthermore, there are many ways to make hydrogen energy. I would like to emphasize that attempts to procure huge amounts of energy while aiming for decarbonization will eventually lead to a hydrogen-energy society.

Hydrogen is the most abundant element in the universe. The sun is also mainly composed of hydrogen, which can be converted to helium, allowing for the creation of a huge amount of energy. This is nuclear fusion. It is solar energy. Gaia uses it, converting it into various usable energies such as coal and oil. We are developing nuclear fusion technologies with their own intellectual abilities. We are truly the descendants of Prometheus. After all, when hydrogen and oxygen are combined, a huge amount of energy is generated, and only water is produced as a by-product. There is plenty of hydrogen in seawater. If this hydrogen can be extracted, humanity can be completely free from complex energy problems.

So, the question remains, how do we secure this hydrogen? Hydrogen energy itself is not an energy source. It is a new energy carrier produced using nuclear power, renewable energy, and fossil fuels. It's like a second source of electrical energy. If water is obtained by boiling water using nuclear energy, it is called pink hydrogen, and if hydrogen is produced by water electrolysis using electricity from renewable energy sources, it is called green hydrogen. It can also be obtained by reforming and extracting it from gas. This is called grey hydrogen. However, in this case, the carbon component is also released along with hydrogen, so it's meaningless from a greenhouse gas point of view. When using CCS technology that collects the extracted carbon components and stores them underground, it is called blue hydrogen.

However, it is impossible to satisfy the fuel needs of Korea, an energy glutton, using only nuclear power or renewable energy. Efforts to meet NDCs are inevitable. In the end, we must use blue hydrogen by utilizing fossil fuels that are sufficiently available. At a certain point in time, price will not be an issue. Energy is needed, so it is necessary to secure small-scale decarbonized energy without requests. Thus, producing blue hydrogen using imported gas is inevitable for Korea which has no energy source. Currently, CCS technology and landfill storage are needed, but if there is a serious conflict between climate response and energy supply and demand, blue hydrogen energy is inevitable. This is because the energy that Gaia can provide humanity is extremely limited.

<수소의 색깔 (The Colours of Hydrogen) >
Since hydrogen energy reacts with oxygen to generate electricity, heat, and water, it is a core energy source that has recently gained popularity among various technologies for greenhouse gas reduction. To use hydrogen as an energy source, it must first be produced. The environmental

friendliness of hydrogen energy is determined by the colour assigned based on the hydrogen production method. It can be any of the following: Brown hydrogen, grey hydrogen, blue hydrogen, pink hydrogen, and green hydrogen.

Brown hydrogen is extracted by gasifying coal (lignite), and grey hydrogen is extracted from the reaction of methane, the main component of natural gas, with water vapour. The advantage is that you can obtain large amounts of hydrogen, but the disadvantage is that it emits large amounts of greenhouse gases. Blue hydrogen, like grey hydrogen, uses natural gas but combines carbon dioxide generated during the production process with carbon dioxide capture, storage, and utilization technology to reduce carbon dioxide emissions by up to 60%. Pink hydrogen refers to hydrogen made by the electrolysis of water with electricity produced through nuclear power generation. Finally, green hydrogen refers to hydrogen extracted through the electrolysis of water using electricity obtained from renewable energy sources such as solar and wind power.

The ultimate production method is green hydrogen, but since production costs are still very high and a small amount of hydrogen is produced compared to the power input, the share of total hydrogen is not significant. Therefore, recently, pink hydrogen extracted using nuclear power generation is also included in the green hydrogen spectrum. According to the 'Hydrogen Economy Performance and Hydrogen Leading Country Vision' announced by the Ministry of Trade, Industry and Energy in October 2021, Korea's goal is to convert 50% or more of hydrogen production in 2030 and 100% or more in 2050 to green and blue hydrogen.

[Cheng, W. and Lee, S., 2022. How Green are the National Hydrogen Fans? Sustainability, 14 (3), p.1930.]

[Ministry of Trade, Industry and Energy, Hydrogen Economic Performance and Hydrogen Leading Country Vision, 2021.10.]

[Energy Policy Communication Centre, Does Hydrogen Have Colours?]

https://eiec.kdi.re.kr/policy/materialView.do?num=218872

https://e-policy.or.kr/info/list.php?admin_mode=read&no=5761&make=&search=&prd_cate=2

There are differentiated attempts in climate response. These include RE100 and CF100 (Carbon Free 100). Both are cooperative programmes on low-carbon energy sources for climate response. RE100 is the centre for renewable energy, and CF100 includes nuclear power and hydrogen. In the case of Korea, CF100 is more suitable from the standpoint of a mix centred on our nuclear power plants. The problem is that there are still absolute companies that must meet RE100 requirements in the global market, so our companies must rely on renewable energy. If Apple and Renault Motors join the CF100, our companies will be able to respond relatively easily, but this is not our choice, but the choice of global companies.

<RE100 (Renewable Energy 100%) >
RE100 is a voluntary campaign to procure 100% of the electricity required by companies through renewable energy sources such as solar, wind, hydropower, and geothermal power by 2050. It was launched in 2014 by The Climate Group, a multinational non-profit company in London, England. The Carbon Disclosure Project (CDP) was jointly launched in 2014 and began as a campaign to support the Paris Agreement adopted by the 21st General Meeting of the Parties to the United Nations Convention on Climate Change in 2015. Companies participating in RE100 aim to achieve 100% by 2050 and have been urged to achieve results of 60% by 2030 and 90% or more by 2040. Currently, more than 380 companies are participating in RE100, including global companies such as Google, Apple, and Microsoft, and domestic companies such as Samsung Electronics, Hyundai Motors, and SK Hynix. According to the RE100 Annual Disclosure Report, more than 60 companies including Google and Apple have achieved RE100 to date, and the average target achievement year for participating companies is 2031.

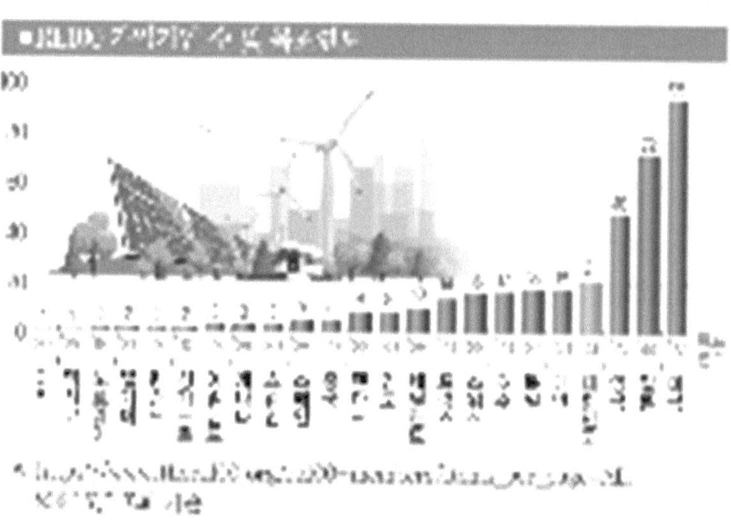

<RE100 Number of Registered Firms [E2News]>

[RE100 official website, https://www.there100.org/]

[2050 Carbon Neutrality Committee, What is RE100? 2022.03.]

[E2News, summary of RE100 target year, 2022.08.]

Source- https://www.korea.kr/news/visualNewsView.do?newsId=148899478

https://www.e2news.com/news/articleView.html?idxno=244546

We maintain civilization, industry, society, and our daily lives by utilizing various energies that have their own circumstances every day. However, I must emphasize that they are all children of solar energy. They only differ in terms of concentration from fossil fuels, with renewable energy being introduced as eco-friendly energy.

Chapter 3

Korea's Energy Story

Looking back, it's interesting to compare the situation of South Korea and North Korea after liberation. At that time, the power plants and factories were in North Korea, and underground resources were much more abundant. South Korea, on the other hand, had only minimal power plants and agricultural land. Of course, there were almost no underground resources. Therefore, when North Korea implemented a power outage after liberation, South Korea experienced its first nationwide blackout. The good news was that there were almost no factories, so it was just an inconvenience in everyday life.

Meanwhile, President Lee Seung Mann signed the Korea-US Nuclear Energy Agreement with the United States and began developing technology for the peaceful use of nuclear energy. North Korea's President, Kim Il-sung, also began studying nuclear energy by sending 30 top researchers to the Soviet Union each year. The history of nuclear energy in North and South Korea is just as deep. President Moon Jae-in, who insisted on demineralization by considering the time of the historical accumulation of nuclear energy, overlooked this history. As far as nuclear energy was concerned, both North and South Korea had tearful stories. As a result, North Korea secured an ICBM-class nuclear bomb, and Korea reached the point of exporting nuclear power plants. Nuclear energy, which supplies high-density energy, is an independent and essential energy source in a country such as Korea that has no existing resources and focuses on the heavy chemical industry. Furthermore, abandoning nuclear energy in terms of the possibility of export industrialization is foolish. However, the safety of nuclear power plants and the issue of spent nuclear fuel are still the most important issues. Nevertheless, nuclear power is still unavoidable.

A long time ago, half of our fuel was firewood. Wood was made in the mountains and used as fuel in each household. However, as coal mines developed mainly in the Gangwon Province, we were finally able to use briquettes. An innovative product was supplied under the name Gugongtan. Every winter, each household stocked up on briquettes in advance. Families had to wake up once at dawn to change their briquettes every winter night. In the 70s alone, schools at various levels received lignite and started fires in stoves during the winter season.

However, in the 80s, when city gas became a reality, the tragedy of dying from briquette poisoning began to disappear. The rapid spread of city gas was a marvellous thing. It was conducted like a military operation. Since the country was under a military regime at the time, it was possible. The oil crisis that occurred during this time was a huge shock to Korea, an emerging developing country. Due to our good fortune, factories had not yet been built, but the trauma led the country to replace 70% of its Petro electric power plants with coal power plants in one fell swoop. This, too, was handled like a military operation.

Korea fostered the oil, electric power, and gas industries, which were the main energy sources, as public property. Although the petroleum sector was successfully liberalized in 1997, electricity and gas remained in a very neutral state despite discussions on restructuring. However, after the restructuring, Korea's energy policy lost its own roadmap and drifted somewhat. However, we have now entered an era where the focus is on the public issue of climate response rather than privatization or competition under neoliberalism; a completely new approach is required. We need to consider a new approach to reconcile the public challenges of increasing efficiency and climate response. I am hopeful that viable alternatives will be identified soon.

However, if I had to pick just one characteristic of Korea's current energy system, it would be its highest energy density in the world. The energy density per unit area (power generation, consumption, etc.) is the highest in human history. However, related technology is a mature field of study, and the level of technology used by power companies around the world is very flat and similar. If you compare it to a submarine, the performance is the same. Doesn't it feel like something is at stake? After all, we are a proud country that has built such a sophisticated energy civilization. Recently, we achieved global prominence due to music groups such as BTS, international film festivals, etc., but Korea is already number one in the world in terms of energy density. Grandpa Dan-gun will be amazed by our achievements.

However, the problem is that even though it is an energy glutton, it has almost no energy of its own. Almost all its required energy needs (oil, coal, gas, uranium, etc.,) must be imported.

Since the international energy market has been stable over the past few decades, it has been possible to import energy with relative stability and at affordable prices. The country manufactured products with imported energy, exported them, and imported more energy with the dollars obtained from exports. If energy prices skyrocket, as seen in Ukraine, Korea's foreign exchange reserves will be rapidly depleted. Only countries with resources are stable in times of crisis. Korea lacks original technology in advanced fields, underground resources and crisis awareness in the economic sector. This is a major cause for concern.

There is a high probability that the price of energy, especially oil and coal, which are traditional energies, will temporarily rise when investment is reduced in the climate response process. Since consumption levels are the same while the amount of energy being mined is small since it is a scarce resource, a temporary price increase is inevitable. The cost burden of energy income will increase whether we innovate the energy system or not. The second IMF can approach the energy crisis in this way. However, everything has gone haywire due to the crisis in Ukraine.

If offshore transportation routes for oil or gas are blocked, how long can our economy survive? We don't have the military power to protect our maritime transport routes. Thus, when events such as China's invasion of Taiwan, the US-Japan-China naval battle, and the war in Hormuz occurred, the power of the dollar was lost, and our power plants came to a standstill. Of course, nuclear power plants and solar power can generate a substantial amount of electricity, but they fall far short of meeting the overall demand. We are a country without resources.

Even if the introduction of resources from outside is smooth, the domestic depletion of position is a realistic crisis. Advances in electrification are inevitable due to technological progress and the pressure of de-fossilized fuels. Therefore, we must continue to build eco-friendly power plants. Nuclear power plants must also be built, and renewable energy must continue to be distributed. Electric charging stations and hydrogen charging stations must also be built. Substations and transmission towers must also be built. However, contrary to what the locals believe, the construction of such public national infrastructure has been delayed. Even now, social conflicts stemming from energy are occurring across the country. Energy plays a major role in our existence as a conflict republic. It's true that our energy density is excessive in the eyes of the locals who are foolish. The depletion of position is a serious obstacle. Who is willing to easily make concessions? Before being ruined by climate change in 2050, we will be ruined by the depletion of our power grid.

Our country has achieved tremendous results. A long time ago, we became the world's highest-density energy society. Because of this, the manufacturing industry also achieved tremendous results, supplying Europe's most advanced weapons. However, to maintain this success, a crisis from within is occurring, and a completely new and fearsome crisis is coming from the outside as well. The warning from Seung Seung Bok's grandson is constantly invading my mind.

Chapter 4

Let's Learn About the Climate Agreement

Everyone is familiar with climate change now. The global climate system is rapidly changing due to greenhouse gases such as carbon dioxide and methane emitted by humans after the Industrial Revolution. Of course, the Earth's climate has fluctuated innumerably over 4.5 billion years. There were times when it rained for 2 million years, and there was also a time when the whole earth was an ice ball. The last few million years have been identified as the Ice Age – a very cold season with glacial periods.

As the Earth entered the interglacial period about 10,000 years ago, it became warmer, the glaciers melted, and the overall sea level rose. The coastline became fixed around 7,000 years ago. Finally, we humans began farming, entering the age of civilization. Then, 200 years ago, the Industrial Revolution began to change the atmospheric composition with the use of coal and petroleum. Of course, the corals of Daeriku Peak and the giant plants at one time had changed the composition of the atmosphere. However, homo sapiens are the only species that caused rapid changes such as a one-degree rise in atmospheric temperature in just 200 years. Gradually, disasters associated with climate change became visible. We often hear about these climate disasters on the news.

The Climate Change Convention (UNFCCC) in Rio, Brazil, had already been signed in 1992. However, over the past 40 years, tangible results have been minimal, and climate disasters have only increased. Even now, the glaciers in the Alps are melting, and the sea level is rising steadily. Through a complicated process, we came to the Paris Agreement, in which all countries participated voluntarily. As a result, each country submitted an NDC. However, the IEA predicts that the global energy demand will continue to increase despite warnings from the IPCC. In other words, it strongly suggests that humanity will eventually find it difficult to achieve the 1.5-degree scenario.

<유엔 기후변화협약과 파리협정 >
The United Nations Framework Convention on Climate Change (UNFCCC) was adopted at the United Nations Environment and Development Conference held in Rio de Janeiro, Brazil, in May 1992. The basic principle of the UNFCCC is a shared but differentiated responsibility, where all parties (Annexes), including developing countries, participate to respond to global warming, and developed countries with historical responsibility undertake differentiated responsibilities. The highest decision-making body is the Conference of Parties (COP), which has been held every year since the first COP was held in Berlin, Germany in 1995. The 27th COP was held in Sharm El Sheikh, Egypt, in November 2022. Currently, 196 countries are participating, and Korea joined for the 47th time in December 1993.

At the third COP held in Kyoto, Japan, in 1997, the Kyoto Protocol (Kyoto Protocol), which stipulated the greenhouse gas reduction obligations of developed countries, was adopted, and came into effect in February 2005, with the planning period running from 2008 to 2020. The Kyoto Protocol introduced emission trading called the Kyoto Mechanism, a clean development mechanism, and a joint implementation system to ratify the protocol. As the Kyoto Protocol expired in 2020, negotiations on a new climate system after 2020 began, and the Paris Agreement (Paris Agreement), which forms the basis for a new climate system in which all countries participate in greenhouse gas reduction, was adopted at the 21st COP held in Paris, France in 2015. The Kyoto Protocol imposes greenhouse gas reduction obligations only on developed countries, but the Paris Agreement is significant because it is a universal system in which all parties participate. Under the Paris Agreement, each country must submit an NDC, which is a voluntary reduction target. As of February 2023, 194 parties have submitted NDCs.

[Ministry of Foreign Affairs, Climate Change Negotiations, 2023.02.06.]

[Ministry of Environment, Paris Agreement Viewed Together, 2022.03.]

Source:https://www.mofa.go.kr/www/wpge/m_20150/contents.do

https://me.go.kr/home/file/readDownloadFile2.do?fileId=236160&fileSeq=1&fileName=f9385cd642b1a043cd78243888378eaac441bc9dc03ac9a64126692e476397d8&openYn=Y

<Nationally Determined Contributions, NDC>

National Decision Contribution (NDC) is a national greenhouse gas reduction target determined and submitted by oneself in accordance with the Paris Agreement adopted at the 21st COP in 2015 as an interim goal for achieving carbon neutrality by 2050. NDC ensures that each country can set reduction targets autonomously and agrees to abide by the principle of progress over time, which reinforces goals over time. In other words, the reduction target of the NDC submitted for the second time must be a target that is stronger than the target submitted for the first. As of February 2023, a total of 194 parties have submitted NDCs at least once.

NDC reduction targets can be divided into 1) Absolute volume (setting reduction targets compared to the base year), 2) Emission forecasts (business-as-youths, BAU), 3) Intensity (setting reduction targets for greenhouse gas intensity in the base year), and 4) Policy and instrument targets. Looking at NDCs submitted by major countries, the EU is aiming for a reduction of at least 55% compared to 1990 in 2030, the United Kingdom is aiming for a reduction of 68% compared to 1990, and the US is aiming for a 50-52% reduction compared to 2005. China has submitted an intensity target for greenhouse gas emissions divided by GDP, and it aims to reduce greenhouse gas emissions by 65% compared to 2005 by 2030. Korea has set a target of at least a 35% reduction compared to 2018 and has recently raised the target to a 40% reduction.

[Carbon Neutrality Green Growth Committee, 2030 National Greenhouse Gas Reduction Targets]

[UNFCCC, NDC Registry, 2022.02.05. search] [Ministry of Environment, Paris Agreement viewed together, 2022.03.]
Source- https://2050cnc.go.kr/base/contents/view?contentsNo=11&menuLevel=2&menuNo=13 https://unfccc.int/NDCREG?gclid=CjwKCAiAxP2eBhBiEiwA5puhNcWCuxFZPMb2gYerdnOpLXliZTJG7j-Z88U3F3hx99BWbzl44-xbExoCQ8gQAvD_BwE https://me.go.kr/home/file/readDownloadFile2.do?fileId=236160&fileSeq=1&fileName=f9385cd642b1a043cd78243888378eaac441bc9dc03ac9a64126692e476397d8&openYn=Y

The final scene in the movie *Tomorrow* is exciting. North American homo sapiens, freezing in the cold, flock to Mexico, where it is relatively warm. The President of the United States implores the Mexican president to develop a border on the condition that all national debt is forgiven. The movie comes to an end when the borders open and the Americans enter Mexico, creating their own happy atmosphere. Perhaps the sequel won't be a war between American immigrants and the Mexicans in Mexico. Imagine the arrival of difficult times with the Second Texas War. Alternatively, the narrative presented in the movie 'Water World' is also highly relevant. While extinction may not be imminent, the prospect of war or even the emergence of humans with gills could become a reality. It is truly disturbing.

But even though the problem is serious, why is the climate agreement that began in 1992 still under-supported more than 30 years later? This is because conflicts between national interests have not been resolved. It is also because when responding to climate agreements, differential changes in competitiveness between countries will inevitably occur. If carbon taxes are levied the same across all areas, there will be no differentiation in the same industry. However, if the imposition of a carbon tax of a few dollars determines export competitiveness, no one can put such a burden on domestic companies first. This is because the factory will close its doors soon. It means an increase or decrease in jobs from country to country, and at the same time, it means large-scale movement of production plants around the world. This is especially true today when interest in attracting factories has grown tremendously.

Let's consider an example to improve our understanding. Currently, Korea has a strong steel company called POSCO. In the global steel market, various steel products compete using similar methods, and every day there is a battle where each product is separated by a few dollars in price. If we impose a carbon tax of a few dollars based on a high level of humanity, POSCO's poor price competitiveness is inevitable. Deciding to reduce the competitiveness of domestic enterprises for the future of humanity is very difficult. Here, a new life and death competition in the steel industry will begin, depending on whether a new decarbonization method called the hydrogen reduction steelmaking method is realized. According to reports from officials, Renault and Nissan require carbon-neutral steel products. Additional investment is unavoidable. It's a huge investment and gamble that risks the future of a company.

However, steel companies around the world are facing this challenge. The climate-related war is already underway.

Of course, to respond to climate change, it is best for the world to adjust the demand for steel itself. It's about reducing automobile purchases and reducing the construction of buildings containing steel bars. Therefore, this IPCC report has begun to emphasize the reduction in demand itself. However, the global demand for steel remains constant. If anything, it will only continue to increase. The demand forecasts of the IEA and the Steel Association are similar. It is the same for other raw materials. Since this means a reduction in capitalism, it is anti-market. It's a very difficult job.

What is our situation regarding climate response? We are completely dependent on trade. We have no choice but to be sensitive to global environments such as the Paris Agreement. Refusing to participate is not possible. We must continue to submit updated NDCs in the future. Even regarding 2050 carbon neutrality, a minimum convincing effort must be clearly shown. Of course, it's not legally binding. However, the goals of the NDC must not be ignored. This is a relatively lax norm. There are only moral accusations against those who fail. The Paris Agreement is such a low-level international agreement. It is not something to worry about in the short term. The Republic of Korea, which has a short-term government of five years, doesn't worry much about conservatives or progress. Therefore, a Progressive Government can easily announce carbon neutrality to the public and suppress energy charges. In the case of the Ministry of Conservatives, there is a tendency to ignore the climate policy by inwardly evaluating it as belonging to the Progressive Group.

In fact, from the political perspective of a five-year short-term government, the 40% reduction in NDC after 10 years is unlikely to be taken seriously. This is probably because the goal wasn't set during the term. That's why they set bold goals.

Right now, the problem is that individual and specific climate standards such as Carbon Border Adjustment Tax (CBAM) and RE100 are operating in the market regardless of the Paris Agreement. The policy of protecting domestic facilities that emit carbon in the same way is not a good idea. The IRA Act implemented by the United States is an attempt to use climate change concerns to return factories around the world to the United States. However, if you look at this issue closely, it has a lot to do with Korea. Even if Samsung or the Hanwha Group build a factory in the US, of course, the ownership belongs to the company. If we don't build a factory in the US, the company will lose its competitiveness. However, there is a stark difference between having a factory in Korea and having a factory in the US. Taxes and jobs in the Republic of Korea will be reduced. Of course, it's better than such companies not being able to go to the US and losing out in the process.

<Carbon Border Adjustment Mechanism, CBAM>
The Carbon Border Adjustment System is a type of trade tariff that the EU intends to introduce to solve the carbon emissions problem. Carbon emissions refer to a phenomenon where such emissions are transferred from countries with strong regulations on carbon emissions to countries with relatively

few regulations. For example, carbon emissions occur when companies move factories located in countries with strong regulations to countries with relatively weak regulations to avoid regulations on carbon emissions. Since the EU has strong regulations on carbon emissions, it aims to pressure other countries to reduce greenhouse gas emissions, strengthen domestic industrial competitiveness, and secure tax revenue through CBAM.

Part of CBAM's draft in June 2021 was revealed to the media. It was included in the legislation (Fit for 55) to achieve the 2030 Climate Goals Plan announced in July 2021. The EU introduced CBAM on a pilot basis on January 1, 2023, and aims to implement it in earnest from January 1, 2026. There are a total of five items affected by CBAM, including steel, aluminium, cement, fertilizer, and electricity. As of 2021, Korea is the fifth largest importer of steel from the EU, which is expected to have a significant impact on our industry and economy.

[Energy Economics Research Institute, (World Nuclear Power Plant Market Insight) Overview and Promotion Status of the Carbon Border Adjustment System, 2022.09.]

[Samyun Kye Corporation, EU Carbon Border Adjustment System, 2021.07.]

Source-https://www.keei.re.kr/web_keei/d_results.nsf/0/CFA267FD5854D1AE492588C6003218ED/$file/WNPMI220923.PDF https://www.pwc.com/kr/ko/publications/research-insights/samilpwc_eu-cbam.pdf

<Inflation Reduction Act, IRA>

US President Biden signed the Inflation Reduction Act (IRA) to support inflation, climate change response, and tax reform on August 16, 2022. IRA aims to reduce household spending on prescription drugs, health care, and energy, create high-paying jobs, and grow the economy centred on the bottom-up middle class. In the clean energy sector, the goal was set to reduce greenhouse gas emissions by 1 Gt by 2030 by reducing environmental pollution, improving clean transportation, and expanding access to clean energy. To support energy security and production in the US, it includes encouraging domestic production of clean energy technologies such as solar, wind, carbon capture, and clean hydrogen, production tax credits for solar panels, wind turbines, batteries, etc., setting guidelines for consumer rebate programmes related to the purchase of heat pumps or energy-efficient appliances, and implementing a tax credit of up to $7,500 when purchasing new electric vehicles.

The impact of the IRA on our companies mainly depends on whether it is produced in the US. If the energy industry, including solar panels, wind turbines, and batteries, manufactures products in the US, our companies can also receive various support benefits such as tax credits. However, in the case of battery manufacturers, if there is even a small amount of minerals or components procured from foreign entities of concern, including Chinese or Russian companies, they are not eligible for the tax credit, so it is necessary for our companies that rely heavily on minerals made in China to receive IRA benefits.

[Korea Industrial Technology Agency, [2022 No. 17] (Issue Focus) US Inflation Reduction Act Comes Into Effect, 2022.09.]

[Plaza International Trade Research Institute, Highlights of the US Inflation Reduction Act and Implications for Our Businesses, 2022.09.]

Source-https://www.kiat.or.kr/front/board/boardContentsView.do?board_id=71&contents_id=eb99f847b77d4d4ab1d485497cd2731e&MenuId=878cb9b6d5ec41bf914ad5c0f590ed14

https://www.leeko.com/upload/news/newsLetter/885/20220829085738177.pdf

How do we go about it? Would it be wise to accept the words of some environmentalists, switch all energy to renewable energy, and abandon energy-intensive industries such as the steel industry and petrochemicals, as the person heading the Ministry of Environment spoke publicly at the discussion? Is it right for the Republic of Korea to take the lead and give up carbon emissions to prevent the extinction of humanity? Some environmentalists think that certain energy-intensive industries such as steel and petrochemicals can be abandoned. However, this is not only an outflow of our national wealth, but also a transfer of emissions overseas, which has no effect on global climate response. Furthermore, will the members of the National Assembly elected from Dangjin, and Pohang agree when their party adopts this opinion?

Is it possible to just emit carbon in a selfish way? This too will be difficult. The international market will be used as a reason to take over Korea's factories. This is particularly fatal to China and Korea, which are manufacturing powerhouses. Nothing is easy in this world. However, there is a group that is very active on this issue. They include the manufacturing industries in Korea. I can confidently say that it is our businessmen who best understand and are willing to resolve this contradiction.

Can we pursue the harmonious pursuit of national interests and human peace? The sense of crisis from a global citizenship perspective blurs the implications for Korea and our response. It is quite possible that it will only generate justifiable discourse and weaken actual response efforts. It is also possible that the government will repeat the mistake of intentionally neglecting rate normalization when it brilliantly declared carbon neutrality in the past. As a result, these humanity-based attitudes are not beneficial to us or our descendants. It is wise to respond strictly by putting national interests first. Only in this way can practicality be further strengthened. Also, if the method is technologically innovative, it goes beyond national interests and is in line with the interests of humanity.

Part 2: Maintain the Status of a Manufacturing Powerhouse as Carbon Neutral

On January 13, 1962, the Republic of Korea announced the First Five-Year Economic Development Plan. Beginning on this day, the Republic of Korea, a miserably poor agricultural country, successfully began the path to industrial sophistication. There's another little-known fact. The First Five-Year Power Development Plan, led by KEPCO, was announced soon after, in the spring of that year. At that time, the government bravely began constructing power plants and transmission towers to supply electricity in advance, keeping in mind the factories proposed in the five-year plan documents that did not yet exist. As a result, Korea's manufacturing and energy industries grew together.

<The First Five-Year Economic Development Plan>

The First Five-Year Economic Development Plan began to be drawn up on May 20, 1961, led by Yoo Won-sik, the chairman of the Finance and Economy Committee of the National Reconstruction Summit, with the participation of some experts including Kim Sung-beom (Industrial Bank), Baek Yong-chan (Industrial Development Committee), Jeong So-young (Sasebuk of the Ministry of Finance), Park Hee-beom (Seoul National University), Kwon Hyuk-ro (Army Lieutenant Colonel), Lee Kyung-sik (Bank of Korea), and Jeong Il-hui. The first draft plan was completed on September 15, 1961, by the Economic Planning Institute, combining sector plans and investment plans submitted by various government departments based on the total volume targets and sector-specific targets proposed in the Supreme Assembly, and was approved by the National Reconstruction Supreme Council and published on January 5, 1962.

The basic goal of this plan was to modernize the industrial structure and promote the establishment of an independent economy. Six major focus areas were selected: 1) Food self-sufficiency; 2) Laying the foundation for industrial structure upgrading; 3)Achieving $700 million in exports and establishing a foundation for groundbreaking improvement in the balance of payments; 4) Increasing employment and curbing population expansion; 5) Dramatically increasing national income; and 6) Promoting science and management technology and cultivating human resources.

[National Archives of Records, Five-Year Economic Development Plan Seen in Records]

Source: https://theme.archives.go.kr/next/economicDevelopment/primary.do#cite01

Over the past 60 years, the Republic of Korea has achieved the miracle of the Han River, becoming a developing country with the desire of its people to live well. It succeeded in drastic globalization and digitalization in the face of globalization led by the United States. As of 2023, Korea is an advanced country.

However, we are staring at a new crisis. Of course, every day is a crisis, but the future crisis is very deep and is likely to last for a long time. The coronavirus crisis has led to a collapse in real estate. Soon, another crisis will occur. We will make every effort to respond to the new crises when it happens. There are crises such as the war for supremacy between the US and China, the fall in apartment prices, and the ageing population, but most people talk about climate response as the most serious crisis today. The climate crisis is bound to have a huge impact, especially in the case of Korea, where resources are scarce and energy consumption is high. Therefore, even amidst various types of crises, we must do our best to preserve our legacy – our strong manufacturing capability and the energy security that comes with it.

If we protect these two assets, we can proudly rise to the ranks of more stable developed countries. Wouldn't it be possible to achieve something like G7? Amidst all our manufacturing capabilities, there is a manufacturing sector where the public and government must work together. This is especially true for energy-intensive industries related to climate response and climate-related innovative technologies and new markets. The core industries of greening, such as electric vehicles, nuclear power, hydrogen, and smart grids, are already rapidly evolving under pressure from innovation around the world. Whoever wins in this evolution will become a global winner.

As Bill Gates argues, public-private cooperation is important in climate-related competition. This cooperation is a new industrial policy in the new trade era. Since the technologies required for climate response that consumers are willing to buy are not yet in demand, it is necessary for the government to take the lead in creating such demand. This policy demand will be the initial nourishment for innovation in the manufacturing industry. Therefore, the government's key contribution is to guarantee initial market demand. We are already paying subsidies for renewable energy and electric vehicles, etc., and in the case of buildings, strong technology regulations such as ZEB (Zero Energy Building) are creating artificial demand.

In the military, this is called the principle of synergy. A Joint Chief of Staff exists for close cooperation between the Army, Navy, and Air Force. Public-private partnerships require wisdom that enhances the technical competitiveness of our companies by refining the design of such technical regulations and assistance, while simultaneously putting in place domestic demand defence policies so that they can be applied in parallel with existing orders such as the WTO, which are protectionist in nature. This represents the government's competitiveness and a new industrial policy. The whole world has already entered this competition. We must become a winner here and maintain and develop our legacy manufacturing capabilities. Only in this way can we maintain our success and wealth. Energy is at the centre of these efforts. Changes in energy itself and the various technology groups associated with it are changing, competing, and evolving. In the end, everything is reduced to energy issues. Prometheus is still alive.

Chapter 1

Korea is Now Clearly an Advanced Country

There was nothing like Starbucks when I was a college student. It was a time when sake was OB, and soju was full of toads. However, we were not sorry that we were poor when we were young. It was a time when employment was waiting for me as soon as I graduated. Years passed by, and now we live in a country where products are everywhere, the public toilets are the cleanest in the world, and natural carpets of greenery envelop every mountain in the country. Some people opened their eyes one day and became bugs, while we discovered that we were living in a developed country. What really happened in the past 60 years? Will we be able to maintain this level of success? Suddenly, my worries have come to the forefront.

More than 60 years ago, we adopted a military strategy for the military. And it was a success. The foundation for the heavy chemicals industry was laid through the development of dictatorship. The Republic of Korea had won another battle. The five republics boldly attempted an open free market economy system. Licensing restrictions on the import of major commodities were abolished, and Korean companies were able to freely export and import. These measures were successful in combination with a global free trade campaign led by the United States. Openness and globalization have further strengthened our dynamism. What will intellectuals, who cried over movie screen quotas claiming that the Japanese culture would devour us, say now? Where are the people who claimed that the FTA with the US would ruin all of Korea's agriculture? We opened our economy and succeeded with our hard work and luck.

As Peter Zaihan said recently, it is true that Korea has seen the virtue of the flat world led by the United States. Korea's opening-up policy, which came a little too close to gambling, exploded in conjunction with these external proposals. China also grew significantly due to Deng Xiaoping's open policies, and Japan has also shown a trend of taking over American real estate. Therefore, Korea, China, and Japan in Northeast Asia have become the biggest beneficiaries of globalization. However, Japan is going downhill after the Plaza Agreement, and China is currently waging a difficult economic war with the United States.

<Plaza Accord>

G5 Finance Ministers' Meeting held at the Plaza Hotel in New York on September 22, 1985. From left, West Germany's Gerhard Stoltenberg, France's Pierre Beregovois, America's James Baker III, England's Nigel Lawson, and Japan's Finance Minister Takeshita Noboru.

The Plaza Agreement was signed in September 1985 at the Plaza Hotel in New York, by the Finance Ministers of the G5 (US, France, Germany, Japan, UK), who intervened in the foreign exchange market to devalue the US dollar against the Japanese yen and German mark. The US dollar continued to strengthen due to America's political and economic status until the mid-1980s, despite America's large-scale deficit. The United States needed to intervene in the foreign exchange market to prevent the value of its currency from falling as its competitiveness in the international market weakened. Other developed countries faced economic downturns because they had to implement excessive monetary tightening policies to prevent the depreciation of their currency against the US dollar.

As a result, the G5 agreed to jointly intervene in the foreign exchange market to induce a decline in the value of the US dollar. Despite the decline in the value of the US dollar in the two years after the Plaza Accord, the US deficit was not resolved, and it was no longer implemented. The value of the yen rose due to the Plaza Accord, but its growth rate fell sharply as Japan's exports declined, which was one of the direct and indirect causes of the country's lost decade.

[Korea Economy, (Glossary) Plaza Agreement]

[Weekly Joseon, 30th Anniversary of the Prelude Plaza Agreement for the 'Lost Era', 2015.09]
Source- https://dic.hankyung.com/economy/view/?seq=6774 http://weekly.chosun.com/news/articleView.html?idxno=8962

Nevertheless, our country is still doing well. At the time, in addition to semiconductors, LNG ships, and Granger, Korea was exporting K9 self-propelled artillery, FA50 jets, and submarines to foreign countries. Recently, our K2 tank won a victory in Poland against the Leopard 2 tank of Germany, a manufacturing powerhouse in Europe. We are clearly an advanced country and a manufacturing powerhouse.

Here's a brief look at the role played by individual industries in the development of Korea. First, our financial industry has always been the subject of criticism. No loan is given without collateral. There are virtually no resource-related industries domestically. At the same time, there is no sign that all overseas resource development projects will recover after the investment failures of Lee Myung-bak's government. Agriculture, fishing, and animal husbandry depend on subsidies to survive. The service industry depends on the domestic market on such a small scale that it is unable to drive growth. The K-culture business has grown tremendously, but the scale at which it generates wealth is too small to be called an industry.

However, the manufacturing industry is slightly different. Examining the contribution of the manufacturing industry to the Korean economy confirms its importance. The manufacturing industry has been consistently accounting for more than 20%

of our total GDP since the 90s, increasing to 27% in 2021. As of 2021, electrical, electronic and metal products account for 54% of total exports (the mining industry including petrochemicals stands at 85%). It has absolute importance in Korea, which has an export-based growth engine. The rapid recovery of our economy during any crisis situation such as the 1997 IMF, the 2008 foreign exchange crisis, and the recent COVID-19 crisis is due to the competitiveness of the manufacturing industry. As regards employment, our manufacturing industry is higher than the OECD average in the US, Japan, etc. Furthermore, in the 2018 World Manufacturing Competitiveness Index of the United Nations Industrial Development Organization (UNIDO), it ranked third after Germany and China.

When considering production, exports, employment, and competitiveness, the Republic of Korea's ability to maintain and increase national wealth is centred around the manufacturing industry. Even when we look at it from the perspective of 'harmonious transformation' discussed in the second half of this book, synergies can be created when we combine our current strengths with carbon neutrality. Furthermore, it is possible to seize opportunities for a new leap forward, making continuous growth possible. It goes without saying that we must continue to be a manufacturing powerhouse.

<Carbon Neutrality>

<The Effects of Announcing Carbon Neutrality [Climate Action Tracker Webpage]>

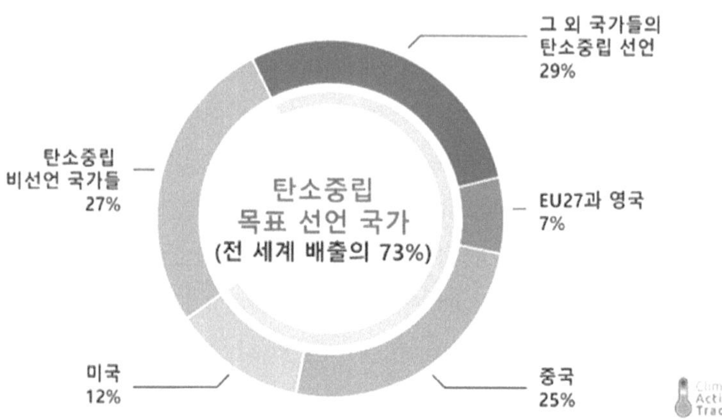

Carbon neutrality means that actual emissions are adjusted to a '0' level by maintaining the amount of carbon artificially emitted by humans to be equal to or less than the carbon emissions naturally absorbed and removed by the Earth. At the 21st Conference of Parties (COP) held in Paris, France, in 2015, the Paris Agreement was adopted where both developed and developing countries participated in greenhouse gas reduction, and an agreement was reached to prevent the rise in global average temperature from exceeding 1.5°C compared to before industrialization. Since then, the Special Report on Global Warming of 1.5°C was published in 2018/10. They recommended that carbon dioxide emissions should be reduced by 45% compared to the 2010 level by 2040 to achieve the temperature target and proposed that carbon neutrality should be achieved by 2050. Since then, many countries around the world have declared carbon neutrality, starting with the EU in 2019. Korea declared in October 2020 that it will achieve carbon neutrality by 2050. Just before the 22nd COP was held in Glasgow, England, in 2022, there were a total of 136 countries that declared carbon neutrality, and the total greenhouse gas emissions of these countries accounted for 88% of the total.

[UNFCCC, A Beginner's Guide to Climate Neutrality, 2021.02.]

[Institute of Energy Economics, [Winter 2021] (Energy Focus) Korea's 2050 Carbon Neutrality Scenario: Contents and Challenges, 2022.01.]

Source-https://unfccc.int/blog/a-beginner-s-guide-to-climate-neutrality?gclid=CjwKCAiAxP2eBhBiEiwA5puhNeYJ4QOQFc3oi3JEV6I9-_5wqkEdPAuKx8ztGGTA7TMDK-5tZMAb_BoC2n4QAvD_BwE

https://www.keei.re.kr/main.nsf/index.html?open&p=%2Fweb_keei%2Fd_results.nsf%2F0%2F40D3CF898F7BCCCD492587CD000CE95A&s=%3Fopendocument%26is_popup%3D1%26menucode%3DS4

Korea is the only country that has achieved both industrialization and democratization after World War II. There are countless historical actors contending for this achievement including workers, businessmen, educators, scientists, administrative officials, etc. It can be argued that energy also wants to brag about its importance, given that the supply of cheap and stable energy was at the core of industrial development. The failure of resource diplomacy promoted during Lee Myung-bak's administration is still painful. Despite this, the energy policy has successfully carried out its mission. Both industry and energy were born simultaneously in Korea.

The success of the energy industry was also due to government intervention through the planning and implementation of comprehensive measures. Regarding the Electricity Supply and Demand Plan, both the 1962 and 2022 plans are essentially the same in terms of their establishment. The elites predicted future electricity demand and built power plants and power grids accordingly in a cost-effective manner. The saga of such elites has been shaken in many ways recently. The times have become too complex, with constant dynamic change. With the advent of smartphones, the intelligence of Director Jang Sam became powerful, and the strength of residents grew due to the tremendous local support. Elected officials are perpetuating confusion by meddling in major plans. For various reasons, questions about the utility of the central planned economy system began to emerge.

The successful practices of the past 60 years are no longer working. I heard a story when I visited the Japan Institute of Energy Economics 10 years ago. In response to a question about the mix adjustment method, the researchers said, “In Japan, when a government official requests mix adjustments based on a basic energy plan, the market accepts it.” Considering the detailed adjustments made by Korea through the sub-legal plans of the Energy Master Plan, it involves a substantial exercise of authority. The fact that businesses fiercely competing in the Japanese market simply obey the administrative guidance of the bureaucracy is reminiscent of our development dictatorship in the past. The administrative discretion exercised by these Japanese bureaucrats has become an abuse of authority by the Korean Government. Right now, for some reason, South Korea is starting to feel overwhelmed by an unfamiliar environment. It is clearly time for a new order.

Chapter 2

A New Huge Crises is Around the Corner

It is true that South Korea can clearly be recognized as a developed country. The big question is, can we maintain this wealth we have accumulated? The favourable conditions have begun to fundamentally change, and if you look at the details of the crisis, the conditions are unfavourable to our manufacturing capacity. This is a crisis.

Many internal factors threaten our success including the fear of ageing, which is difficult to handle at the age of 100, the highest suicide rate in the OECD due to extreme hardship, the world's lowest fertility rate, and a population cliff which is right around the corner. Some research institutes say that our nation will be the first to disappear. There are countless crises we are facing now and will continue to face, such as the diversification and deepening of the varied conflict structures including progressiveness and conservatism, different age groups, regions, classes, and industries. In this situation, if one of Samsung's semiconductors loses its competitiveness, it is inevitable that our economy will face a crisis. In particular, the recent collapse of the real estate bubble, the hopelessness of gap investors, and concerns about the series of bankruptcies of construction companies are being compared to Japanese precedents, and a serious sense of crisis is pervading Korea as a whole.

Here, I would like to raise an issue that is often overlooked. It is a problem of depletion of space to the extent that additional infrastructure cannot be built. The energy sector still needs more infrastructure. Even after building the expensive power plant on the East Coast, the construction of transmission towers was delayed due to conflicts with residents. It is impossible to know whether power generation will begin. There's no space to dump wastes, and hundreds of garbage piles are scattered all over Mount Kumsugang. The dispute over whether to generate nuclear power is intensifying, but spent nuclear fuel, which completely ignores maintenance and progress, is also unable to find a resting place. Offshore wind power, which is a renewable energy source, is greatly disappointing due to conflicts with fishermen. This extreme rhetoric raises fundamental questions about Korea's future sustainability. There are no concessions for the community.

In the face of a looming crisis, can we break through in a way that exerts collective energy like we did in the past? In the event of a crisis, will a miracle such as saving money happen again? I'm proud that my generation which received nationalist education, lived by making concessions to their individual lives, little by little, while insisting on good fortune. However, we have now become a society where the existence of individuals is recognized as more important than the existence of nations. For each of us, the risk of unemployment or loss of promotion that may happen to us right away or within a week is a more obvious and serious risk than this huge crisis. The crisis between me and my family is the only real crisis. What is clear, however, is that the long-term misfortune of each of us will converge with the overall level of wealth of our Republic of Korea. National interests will ultimately align with our individual interests.

These internal difficulties are, however, something we must work hard to overcome. The new crisis we must discuss is attacking us like a tsunami from the outside. It's literally something that's new and huge. Therefore, a completely new response is needed. The phrase 'victory or disobedience' in the art of war has never been more real than it is today. It's more dangerous! The past 60 years have been favourable, but now these good times have come to an end. Various claims that we have now become a developed country, while our consciousness and perception are at the level of a developing country, are meaningless. This is because it can all come crashing down. Our place in the world as a developed country is yet to be confirmed. It's very much at stake.

In particular, the outsider case is very impressive. The competition for supremacy between the US and China is forcing us to make difficult choices. China is not as daunting as one might think. The recent US-led supply chain restructuring is a major crisis for us. In the end, it will sanction the role of global factories allowed to China in the global supply chain and at the same time prevent the outflow and sharing of advanced technology. I wonder what advice Kwanghae-gun, who struggled amidst the conflict between Ming and Cheong, would give as today's situation unfolds. The decline in the liberal trade structure is also a major crisis for countries that live and eat on exports like us. US President Biden blatantly says that only domestic production technology and products are allowed to build domestic infrastructure. As seen in the Ukraine crisis, large-scale wars have also become an ordinary event. These changes in conditions are fatal for those of us who eat and live through exports.

Here, we need to consider the issue of energy supply and demand. We have built the highest high-density energy society in human history since Grandpa Dan-gun founded the country. However, there are also aspects that are somewhat uneasy for a country with no underground resources to be proud of. Securing resources was relatively easy under the free trade system. In other words, if the factory worked well and earned foreign currency, I could import as much energy as I wanted from the international energy market. This has worked just fine over the past few decades. However, the period of comfortably importing oil and gas from these markets has now passed. The war in Ukraine is proof of that.

<Change of Energy Usage>
If you look at energy usage per capita from 1965 to 2021 in major countries, some developed countries, including the United States, peaked in the past and showed a recent decline through energy efficiency, etc. On the other hand, energy usage in Korea and China increased by leaps and bounds as their economies grew. Korea's energy consumption per capita in 2021 was 67,397 kWh, which is the highest among the OECD countries excluding Canada, Norway, and the United States.

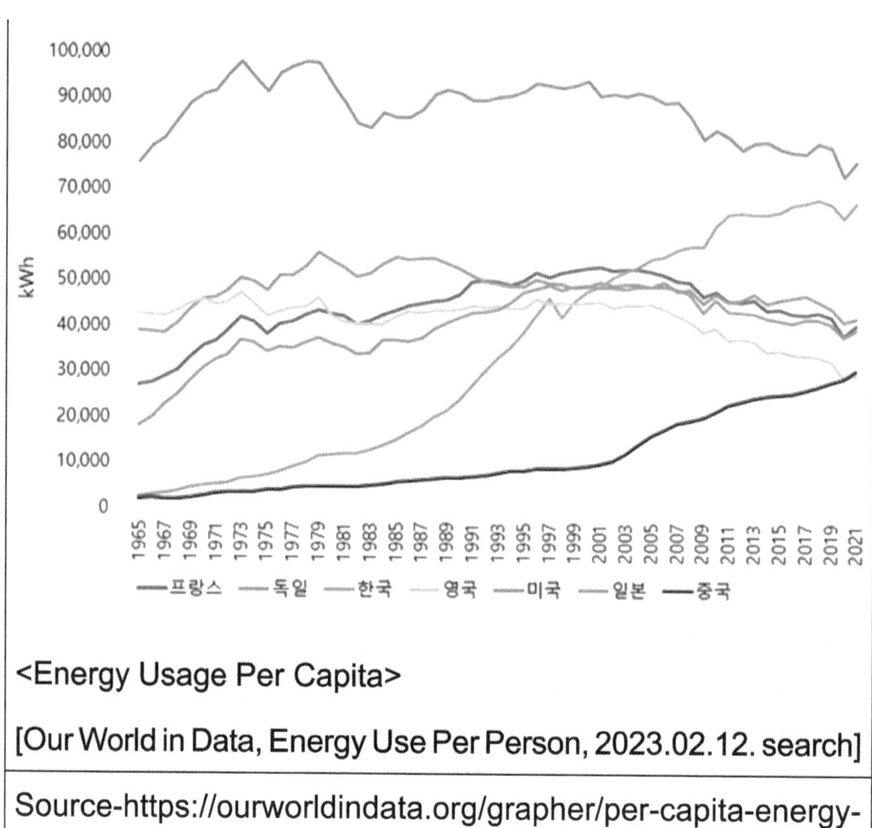

<Energy Usage Per Capita>

[Our World in Data, Energy Use Per Person, 2023.02.12. search]

Source-https://ourworldindata.org/grapher/per-capita-energy-use?tab=chart

For example, if a war breaks out between the US and Iran in the Strait of Hormuz, our maritime trade routes will immediately be threatened. If the supply of oil or gas is delayed, a certain level of reserves may cover our needs, but there is a limit. If a fluctuation in the volume of energy occurs, our manufacturing industry will immediately stop, and all economic activities will come to a stop immediately in Korea, which has managed to build the world's highest-density energy society without regional cooperation related to energy. If electrical energy is cut off beyond a factory's fuel shortage, a tragedy that cannot be compared to the first and second oil crises will occur. Perhaps there is a great possibility that the US will continue to demand stronger and more frequent changes in our supply chain due to its blockade policy against China. It is impossible to predict what will happen to our supply chain if a war broke out between the US and China over Taiwan. Didn't the shortage of urea water products cause great turmoil?

However, a real crisis greater than this awaits us, and it is climate change. It is going to be a struggle for Korea to sustain the world's highest-density energy society. Our dependence on oil is absolute. Even when it comes to supply and demand, the country is unprotected against climate regulations. Furthermore, domestic resources are extremely scarce. Using nuclear power and renewable energy, which are domestic technical resources, is like forcing Kang Ho-dong to eat a vegetarian diet. Private-level RE100 and European CBAM, which block exports, will gradually increase their influence. Our success in entering the Chinese chemical industry, which was the source of our wealth, is just a mirage now. Climate change is a fundamental crisis that will alter the status quo.

What is the status of developed countries overseas? The United States believes that globalization promoted in the past made its blue-collar people poor. This is statistically true. Unexpectedly, the US Democratic Party's solution, which insisted on an open policy, was the Inflation Reduction Act (IRA). The IRA, which is extremely protectionist, forcibly attracts factories that exist overseas to the country. It is also part of the 'Live Well Alone' campaign that originated in the United States. The United States hopes to attract key greening technologies and factories to its domestic supply chain through this. It can be understood that such a radical approach is employed due to the overwhelming influence of blue-collar workers in US politics.

The EU identified the IRA as a discriminatory protectionist measure and called for amendments. However, in the end, they ended up promoting the European IRA, the Carbon Neutral Industry Act. Presently, the principle of free trade such as the WTO seems to be disappearing only in climate-related industries. Countermeasures to sustain the competitiveness of our related industries are inevitable. This is because many IRAs overlap with Korea's main manufacturing capabilities. It's not just electric cars that are currently a hot issue between Korea and the US.

The fluctuations in the volume or price of petroleum, rapid changes in the global supply chain, the pain caused by extreme fine dust, and the distortion of the country due to rising sea levels are all serious issues. This means that our strategy must completely change. Considering that recent academic energy security concepts also emphasize resilience after an accident, it is reasonable to view supply and demand unease or resource security failures as constants. In other words, long-term cost-effective supply and demand stability is likely a thing of the past. We must get rid of old habits. The principles of past successes may now lead to failures. Here, I am reminded of Chairman Lee Kun-hee, who used to ask his wife to change everything except his children. Over the past 60 years, we have used energy to solidify our successes. Today, the success of energy can also possibly lead to future national failures.

Chapter 3

The Public and Private Sectors Must Work Together in Certain Manufacturing Industries

Preserving the most useful and powerful manufacturing capabilities is the most important strategic choice for our future. We have secured significant global competitiveness in the fields of semiconductors, bio, defence, information and communication, construction, consumer electronics, steel, nuclear power, hydrogen, and automobiles. Also, recent activities in the fields of artificial intelligence and aerospace stand out. These competencies will sustain our wealth in the future.

However, it is important to recognize that the characteristics of each manufacturing capability is quite different. In the case of an industrial group represented digitally, it is strictly democratized. For example, what is the government's role in strengthening the competitiveness of Samsung Electronics' semiconductors? Government support for research and development is likely vital for progress in various domains. Tax benefits may be helpful due to President Yoon Suk-yeol's strong will, but they won't be fundamental. Rather, solving trade regulation issues such as RE100 will be better appreciated from Samsung Electronics' point of view. This is because it's not something Samsung Electronics can do on its own. There are individual situations where independent efforts alone will provide only limited results in each of our main industries. Thus, a more strategic collaboration between the public and private sectors is essential.

On the other hand, there are also industries where the government must actively take the lead. The most apt example is the energy sector, particularly, building a hydrogen society, which is a new option. Compared to other energy sources, hydrogen boasts a high density of nutritional value. Therefore, hydrogen is suitable as an energy source for high-density manufacturing powerhouses. There are realistic restrictions on the total amount of nuclear power and renewable energy that can be supplied to Korea. After all, the use of fossil fuels is inevitable. In the end, large-scale blue hydrogen, which is a technology that uses the decarbonization of fossil fuels, is also inevitable for climate response and the coexistence of fossil fuels. Hydrogen is also compatible with nuclear power plants and renewable energy. However, active government intervention is inevitable in the development of advanced knowledge such as the supply, transfer, and utilization of hydrogen. The approach is fundamentally different from the semiconductor support policy. In general, greening requires corporate innovation efforts to be carried out in parallel under the leadership of the government.

Since Lee Myung-bak's government under the banner of carbon neutrality and green growth in Korea, each government has been making its own systematic efforts regardless of performance. However, it is true that the greening of our industry has been slow. Despite setting NDC goals, due to practicality issues, methods of the past are still being used. Carbon neutrality is still in the planning stage and is not yet an actual policy. Climate policy and NDC are seen as environmental agendas. However, Moon Jae-in's government, which actively pursued this issue, only presented a stylistic vision rather than bringing about actual change. Concrete measures such as fee adjustments have been suspended. If greening fails, Samsung Electronics or POSCO will have to move their factories overseas. In this respect, investment in energy security and the greening of the manufacturing industry necessitates closer public-private collaboration under the leadership of the government.

Korean manufacturing companies enthusiastically responded to the carbon neutrality declaration. SK, POSCO, Doosan, Hanwha, etc., and even Samsung declared RE100 at the end of last year. Our main manufacturing industries also have a very strong desire to turn green. Most of the greening projects, such as electric vehicles, hydrogen, and offshore wind power, operate on energy systems, and their results are traded. Therefore, the government must proactively innovate the energy system to accept their technology. The future of Korea's competitiveness depends on how quickly, technologically and innovatively these integrated transitions are implemented. With a strong manufacturing industry, our energy security will be stable. In other words, industrial and energy policies must converge and innovate in an integrated manner. They must also be processed faster, and more comprehensively than other countries, just like racing. That's why it's called 'Green Racing.'

Korea has more success stories through such collaboration than any other country. An example of linking an energy policy with an industrial policy is the efficiency rating system. It is a system that is already well known for its Grade 5 mark that consumers usually see when they buy home appliances. This system was urgently implemented in 1992 by the Ministry of Energy. I remember vividly that the consumer electronics industry fiercely opposed this system at the time. The Ministry of Commerce and Industry, which oversaw such appliances, also expressed active opposition. A corporate official asserted that American Whirlpool products would dominate the Korean market when this system was introduced. However, after the implementation of this system, Korean consumer electronics achieved world-class efficiency through systematic competition created by domestic demand. This became the cornerstone for securing global competitiveness with regard to Korea's white appliances. This is also a testimony to the success of related companies. The technical regulation of energy efficiency not only improves the efficiency of the market but also increases the competitiveness of related companies.

In a country without resources like South Korea, the most important strategy for securing resources is to arm ourselves with technical resources. Indeed, the Club of Rome, in their report 'The Limits to Growth,' expressed reservations about the role of technology. However, at this point, Korea's only solution is technology. Technological assets cover not only energy sources, but also all application areas such as electric vehicles, hydrogen vehicles, smart grids, and hydrogen reduction steelmaking. We are competitive in all these industries. We can talk confidently to the Club of Rome. The Republic of Korea affirms the role of technology in ensuring sustainability. This is a new industrial policy where the government must take the lead in the future.

Our companies are already actively investing. On the other hand, the Central and local governments are too helpless. In the past, the Moon Jae-in government was highly motivated, while the Yoon Seok-yeol government lacked interest. Solar power is flocking to red waste energy, and there is too little interest in future food such as offshore wind power and hydrogen energy. The profits obtained from the direct introduction of gas will be arbitrarily recovered based on the judgement that it is an unfair windfall. Green racing is not a competition between companies, but a competition related to the ability of public-private partnerships between countries. It's not a game that companies can play alone. In that respect, climate-related green racing is a battle of the sum of national forces. But we have made no progress in this matter. We are already losing the race against time.

Chapter 4

Why is Carbon Neutrality a Core Policy of Manufacturing Powerhouses?

We must protect manufacturing capacity and energy security at any cost amidst turbulent conditions. If so, how do we keep these two parameters healthy and develop them? What policies we put in place at the Centre is very important. Which of the many government policies to use as the central policy (anchor) should be the core of our national strategy? However, the major policies are as numerous and complex as the number of government departments and committees. Furthermore, due to the nature of the five-year short-term regime, policies change on a whim. Given the reality that the national tone changes every five years, which policy will receive sympathy from the public for the medium to long-term challenge of green racing, guaranteeing long-term innovation?

Since 1962, the government of the Republic of Korea has led the country by formulating long-term plans. Excellent government officials have worked patriotically and have made strides with companies to this day. Numerous plans, such as the five-year economic development plan and the basic energy plan, have been successfully implemented. The power that led to the development was the result of both the public and private sectors actively accepting and working to achieve the goal of 10 billion dollars in exports.

However, gradually, the government's plans became too diverse and complicated, and issues were getting tangled up. Simultaneously, the policy stability was greatly undermined by the characteristics of the five-year regime. There is some confusion as regards the traditional leadership of Korea. This confusion is likely to continue even if the regime changes. Nowadays, this instability is perpetuated as elected positions in the National Assembly and regional and local governments, which are renewed every four or five years, completely overwhelm bureaucrats who have passed the exams. Popularity has become more important than skill.

However, this phenomenon may not be unique to us. Today, even Director Jang Sam can obtain a huge amount of information through smartphones and see through the world. A book that is popular in Europe presently is *The Age of Ungovernability*. Overcoming the limitations of the monarchy, an elite whose qualifications have been verified, that is, a bureaucratic society governed by a small number of government officials, is losing its absolute advantage in terms of intelligence. Bureaucrats, under the separation of powers, who must make and execute plans, struggle to achieve a balance between pleasing ruthless elected offices and smart citizens. Achieving this balance is burdensome.

In such a complex world, we must share a common direction where something is central. Traditionally, Korean people overcome a crisis by pooling together their strengths when a common goal is set. This is the voluntary dynamism of the people unique to the Republic of Korea. We are a nation that is deeply engaged in its political affairs. What is the central policy that will unite our citizens or our consumers? What issues will contribute to national interests and bring about world peace at the same time? And what policy issues will motivate our companies?

A new effort is needed to stabilize government plans that have become so weak. How can these things be made possible? How can companies trust government policies and invest? A central plan must meet the following conditions: All citizens must be aware of it, it must match the direction of the evolution of future technology, it must be able to cover other plans, it must match the direction of the evolution of our manufacturing industry, and it must be globally bound to long-term goals. This must be our anchor.

I boldly propose a climate change policy of carbon neutrality and NDC which should be our central policy. Even now, these climate policies have a very high status in terms of form, and the Carbon Neutrality Green Growth Committee is working towards implementing them. However, it is lacking, vague, and at the same time, overly rigid and restrictive. In addition, there is a lack of public awareness, consensus, and support. For now, it's just one of the committees.

Of course, carbon neutrality puts a huge burden on our economy considering its feasibility and the magnitude of its costs.

I was impressed by how economists feared that the country would be ruined when the national reduction target was increased from 30% to 36% during the Park Geun-hye administration. It's such a difficult and burdensome task, and it's a nuisance. However, if you look at the spirit of the Paris Agreement, goals are set autonomously, and there is no penalty for failure to achieve them. However, only the set target withdrawal support rules must be complied with. Our government is overly rigid and obsessive about achieving goals. It's enough just to move faster than the other countries.

On the other hand, our manufacturing companies, which are already globalized, have a better understanding of these climate regulations as a rule of thumb compared to citizens and governments. As the reality of such regulation hits us, many companies have recognized it as a new business opportunity. Therefore, they actively accept carbon neutrality. From the public's point of view, I believe climate response will be cooperative because it is almost completely consistent with the fine dust response policy. Most of the implementation tasks of climate policy are healthy, with few side effects, even if their intensity is somewhat excessive. From the perspective of resource-poor countries like ours, these are projects that need to be promoted even more vigorously. It is a medicine that is good for the body even though it is used to combat the climate. Carbon neutrality can be used as a noble and practical goal, uniting all citizens. If we make good use of our strengths, we can sustain our achievements over a long period and simultaneously contribute to human coexistence by achieving innovation. What is needed is wisdom and courage.

Climate and carbon neutrality are poison to us, but they can serve as medicine too. The manufacturing capabilities currently secured must be used as an opportunity to take another leap forward to develop new manufacturing capabilities before the existing ones become obsolete. If we innovate faster than other countries, we can protect our position in the battle for supremacy between the US and China. Furthermore, Korea may become a more powerful country. However, for us to achieve this, it is necessary for our citizens to agree on carbon neutrality. Without consensus on the burden of costs for consumers, the scope may be limited to Germany alone. In addition, since it's a long-term competition, consistent support is needed. Without the political support of the people, Germany will be the only contestant. In conclusion, whether we should maintain our status as a manufacturing powerhouse depends entirely on the people's choice.

Part 3: Carbon Neutrality Breaks Through Innovation in Energy Technology

GREEN RACING

Green racing is a word that was introduced a long time ago. It's also something we've been putting a lot of emphasis on since Lee Myung-bak's government. There was a time when the world was proud of its green growth as a trademark of Korea. However, compared to our past achievements, our performance has been very shabby. Global companies are trying to make stronger voluntary innovations to get ahead in the rat race. This is an age where Renault and Nissan demand carbon-neutral steel from POSCO, just as Apple demands RE100 from Samsung. The race has already started. Everything is decided by who innovates faster.

The essence of green racing is ultimately about preempting the technology necessary for climate response. Who will succeed in making steel using the hydrogen reduction method first? Who will first produce semiconductors that satisfy RE100? Who will be the first to build more economical electric vehicles and charging infrastructure? Who will be the first to preempt the ultimate technology of nuclear fusion? A war of technological power such as this has begun. Competition over which country will develop these innovative energy technologies first, industrialize them first, distribute them first, and export them first is a key element of green racing. And although we don't know it, and the government doesn't realize it, our companies are already in a fierce battle as regards green racing. The race has already begun. The United States is attempting to take not only factories in China but also factories in Korea through the IRA law. If you fall behind, you die.

Until now, industrial policies for the manufacturing industry, low-carbon energy policies, and environmental policies have been poorly interlinked. However, as fusion gradually begins to skirt issues surrounding carbon neutrality, the need to change the role played by energy will become critical. This is because the role of energy in industrial, environmental, and climate policies is being expanded and strengthened immensely. In short, the mission of energy over the past 60 years is going through a fundamental change. From my point of view, considering that I've been working in the energy field for over 30 years, it feels like I am being transferred from a rear military unit to a front-line assault force. I'm excited. A direct example of this is the US IRA corporation, not CBAM or RE100. The mission of energy is changing.

The goal is to maintain the competitiveness of current manufacturing countries along with strong digital capabilities established through green innovation in manufacturing. At the same time, going green necessarily involves moral values, so it secures the morals of developed countries through moral innovation. If this fails, it is very likely that it will not maintain its current status as a developed country and a manufacturing powerhouse. In other words, if you want to survive, you must move and move fast. That's why it's called racing. We must not miss this junction where supply chain reorganization is rapidly changing.

Geopolitically speaking, Korea has a good chance to win. It already has a strong defence industry. It is also possible to provide replacement sites for factories that have left China. Office space in Hong Kong, Singapore, and Tokyo can also be maintained in Seoul or Incheon Songdo. Europe wants to cooperate with Korea as a base for offshore wind power. We have many advantages. In this respect, the core of the KAIST Future Strategy series published annually by KAIST is The Era of Kijeonghak. This is a counterpart of present-day geopolitics. In other words, it emphasizes that supremacy lies not in geographical location but in technical advantage. Specific details, such as core technologies, require some additional discussions. Big tech companies in Silicon Valley (US) have also been shocked by the recent advent of Chat GPT. This is because technology is evolving much faster than we had anticipated. This means that the seven core technologies of the future proposed by Geomatics also have a high probability of fluctuation. The pace of innovation in energy technology is also currently increasing.

The perspective of a new industrial policy that combines energy policies and industrial policies with a focus on the ultimate technology for climate response becomes even more significant. In short, this is the essence of green racing.

The key to green racing is to innovate quickly. We must innovate faster than other companies in other countries. The spirit to innovate quickly is critical. Citizens, consumers, and residents must all sympathize and quickly embrace innovative energy technologies. More subsidies must be offered at an accelerated pace in a shorter period. It is necessary to demonstrate and establish a mass production system more quickly. We must take the lead in reorganizing global supply chains through strategic collaboration with other countries quickly. And we must transform the physical foundations of our buildings, transportation, and industry. The direction is fixed. Now all that remains is whether it will move fast. The quickest country to achieve this will be the winner.

Chapter 1

How to Respond to the Climate Agreement

Let's look at what is needed to respond to the climate agreements we hear countless times daily. Let's also consider what will happen in Korea if we continue with the status quo.

We must adapt to climate change and respond to the climate agreement since we are part of the Paris Agreement. In this regard, there are three key issues to consider. The first is the Declaration of Carbon Neutrality under the Paris Agreement and the Declaration of Position regarding the NDC. Of course, we must stick to the Paris Agreement. However, the Paris Agreement itself is limited in that it is only morally binding. The point is that it's flexible. Therefore, it is necessary to wisely sort out Korea's position regarding a realistic adjustment of the NDC. Since this is currently legally covered by the Carbon Neutrality Green Growth Committee, I won't deal with it in this book.

The second is the problem of adaptation. The climate system is changing constantly. Currently, South Korea does not experience extremely cold temperatures like that of Siberia. Also, the four seasons have now become two seasons, summer and winter. As rainfall changes, changes in crops are inevitable.

Also, over time, the sea level along the Korean Peninsula will rise, and several cities will be submerged. Thus, adaptation is more important than reduction. It is necessary to predict and respond to changes in nature. If we succeed in adapting to climate change and can produce and export high-value crops, it will be a great achievement. It's hard to be optimistic considering the rising sea level. If these adaptations are successful, climate change may be a blessing. It's a good idea to get opinions from experts in the field regarding this adaptation. This book mainly deals with stories about energy.

The third issue deals with matters related to reduction, and the control of medium to long-term emissions such as NDC and carbon neutrality. It's about the government's roadmap for greenhouse gas reduction by sectors, such as household businesses, transportation, industry, and transformation sectors. Korea has already submitted its NDC based on the Paris Agreement. Debates over the appropriateness of the reduction targets continue to be raised. For example, during the Park Geun-hye administration, Korea's reduction target was raised from 30% to 36% before the President visited the United States. The image of government officials reflecting and expressing concerns about the precarious situation of South Korea at that time is commendable. However, since carbon neutrality was declared only a few years later, it is difficult to speculate on the specific thoughts and opinions of government officials who expressed concerns about the situation at that time. After all, even at a considerable cost, we have made a formal commitment to the international community to reduce emissions.

<Nationally Determined Contributions, NDC>
In June 2015, Korea established the first Nationally Determined Contribution (NDC) aiming for a 37% reduction compared to the 2030 Emissions Forecast (BAU) and submitted it to the UN. Under the Paris Agreement, parties are required to establish and notify an NDC every five years that includes goals that are more advanced than existing NDCs. Therefore, the reduced target for December 2019 was changed from the existing emission forecast method to an absolute value method, and a 26.3% reduction target compared to 2018 was submitted to the UN in December 2020. After that, as a follow-up to the 2050 carbon neutrality declaration, they set a goal of reducing emissions by at least 35% compared to 2018, which is the minimum NDC standard specified in the Basic Act on Carbon Neutrality and Green Growth. Korea recently raised the target to 40%. Looking at the reduction targets by sector, the reduction target for the transition sector is the largest at 119.7 million tonnes CO2eq., and the reduction target for the industrial sector is 37.9 million tonnes CO2eq. It also plans to reduce 70.5 million tonnes of CO2 eq through sinks such as forests, carbon dioxide capture, utilization and storage. [Carbon Neutrality Green Growth Committee, 2030 National Greenhouse Gas Reduction Targets] [Relevant Departments Jointly, 2030 National Greenhouse Gas Reduction Target (NDC) Increase Proposal, 2021.10.]
Source-https://2050cnc.go.kr/base/contents/view?contentsNo=11&menuLevel=2&menuNo=13

https://policy.nl.go.kr/search/searchDetail.do?rec_key=SH2_PLC20210275949

More than 20 years before discussions on the climate change agreement began, most discussions were focused on reduction. In terms of negotiations, developing countries were not obligated to reduce emissions under the Kyoto Protocol, so the burden of negotiations on Korea was relatively low. Also, it is reasonable to think that the climate issue was a story that was of no concern to the immediate future, probably because there were few extreme weather events, unlike today. It was an elegant discussion unique to Al Gore. I remember that the issue of adaptation was not a major issue even in discussions held within the government. I think it was of interest to some environmentalists and meteorologists.

Reduction is basically discussed by categorizing it into buildings (homes, businesses, etc.), transportation, industry, and conversion (power, etc.). There was a time when reductions were overshadowed by energy savings. However, as we become more advanced and various social and material bases expand, reduction is turning into an extremely complex systemic issue. It is a huge investment project where a huge material base and assets must be completely replaced while sufficient depreciation remains. For example, what do we do with the hugely expensive coal-fired power plants? Nevertheless, we have promised reductions by committing to the NDC. Furthermore, the Moon Jae-in government officially declared carbon neutrality. Based on the spirit of the Paris Agreement, no retreat from the goals is possible. This reduction process involves huge costs. Currently, if we use innovative domestic technology, it is a welcome catalyst for innovation in our manufacturing capabilities. An enormous amount of money will be converted into an investment for our industry.

Major manufacturing companies in various countries are already participating in voluntary technology regulations unrelated to the Paris Agreement, such as RE100. As a result of these changes, companies in various countries are making efforts to become carbon neutral, and the pace of change is enormous. However, our policies are tending to appear like the fool in the shower. The electric vehicle subsidy in the IRA between Korea and the US is just a precursor to that. Now, energy policies and industrial policies are gradually being integrated. The WTO is not even visible. Despite the crisis in Ukraine, the international community has not abandoned carbon neutrality. Rather, Germany is showing a stronger commitment to carbon neutrality and renewable energy to be free from Russian gas. There may be hurdles, but humanity will move in the direction of carbon neutrality anyway. The possibility of implementing carbon neutrality by 2050 is not important. It is very likely that the tipping point will be exceeded anyway. The problem is that if Korea lags behind its competitors, related industries will be lost. If we innovate faster than our competitors in the market, we will succeed. This is because green racing is a new industrial policy that protects national interests rather than protecting Mother Earth.

<Tipping Point>

A tipping point refers to a critical point where a small change can radically change the state or development of a system. It means a point that causes irreversible changes in the climate system even if global warming is stopped. Climate change is already causing phenomena such as the collapse of ice sheets in Greenland or western Antarctica, the loss of the Amazon rainforest, and the loss of the northern part of permafrost.

McKay et al. (2022), who recently studied tipping points due to climate change, selected 16 potential lists due to global warming. Of these, four, including the collapse of the ice sheet in Greenland or western Antarctica, are likely to reach tipping points even at current levels (1.1 degrees higher than before the Industrial Revolution), and it is estimated that they will almost certainly reach the 1.5-degree increase target agreed upon in the Paris Agreement. Tipping points are difficult to accurately predict, and one change may reinforce another, leading to a domino effect.

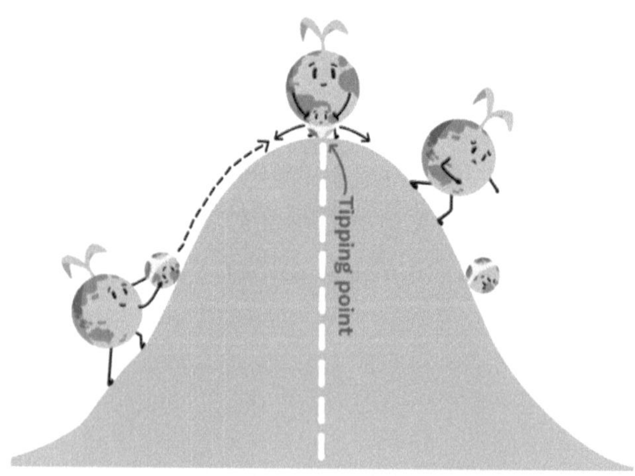

<Tipping Point [Climate Science]>

[Armstrong McKay, David I., et al. "Climate 1.5 C global climate could trigger multiple climate tipping points." Science 377.6611 (2022): eabn7950.]

[Han Kyo-rye, critical point until human extinction, 5 climate crises have already been missed, 2022.09.]

[Climate Science, Tipping Points: Why We Might Not Be Able to Reverse Climate Change, 2022.02.]

Source- https://www.science.org/doi/10.1126/science.abn7950

https://www.hani.co.kr/arti/society/environment/1058469.html

https://climatescience.org/advanced-climate-climate-tipping-points

This approach is a formula that applies equally to every country on the planet. Everyone is intensely thinking and innovating in this sector – countries, companies, and individuals. As I mentioned before, humanity will not be annihilated due to climate change. Winners are bound to exist. In this competition, countries or companies that adapt or innovate quickly will be the winners. The Clinton administration was active with regard to climate response. Climate response is equally a burdensome issue from the standpoint of an existing industry in any country. Citizens are also expected to push back against the cost burden. Clinton had given a speech persuading the public to pay for climate response as it would eventually create millions of new jobs. Al Gore attempted to protect the planet by preaching about the climate crisis, but President Clinton persuaded the people by creating jobs.

In any case, those who adapt and innovate quickly have an advantage. This trend is called green racing. At the root of it all, there is energy. The only solution to maintain strong manufacturing capabilities in the US and China is technological innovation and rapid market transformation. Like Jeju Island's attempts to become carbon-free by 2030, it's a change that remains underdeveloped even though many attempts have been made. Climate action is unavoidable. We're the right size for large-scale innovation and transformation. We should be willing to use our country as a test bed for climate response. Only in this way can we protect our manufacturing industry and maintain our wealth. It is good to focus more on reduction and adaptation.

Pictures related to Jeju CFI

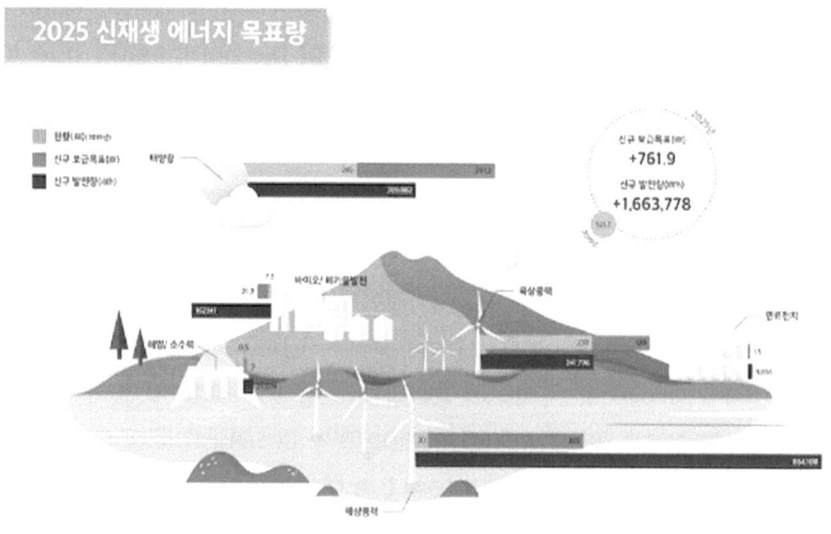

Source-http://www.kea.kr/elechistory/access/ecatalogt.php?call mode=&catimage=&eclang=ko&Dir=564&um=t&start=0

https://www.investkorea.org/jj-kr/bbs/i-1154/detail.do?ntt_sn=3

This is the cover of the November 2016 issue of The Electric Journal. I think it's appropriate, but I think it would be wise to ask the Electric Association and use the original.

Chapter 2

Climate Response and Breakthrough Technologies to Protect Manufacturing Powers

The first expression of concern about the sustainable development of human civilization in modern times was *Limits to Growth* published by the Club of Rome in 1972. However, there is one part of this report that I struggle with. It pertains to the role of technology. Will the technology that is expected to evolve over time solve or worsen the problem? What's interesting is that they deem scientists as 'naive optimists' and are wary of excessive expectations of technology. Currently, as in the movie Kingsman, the usual method is impossible, even if it is an attempt to forcibly reduce the population by half. In the end, it is reasonable to think that technological innovation, especially innovation in energy technology, is the only rational solution.

The beginning of full-scale government-led energy technology began with the oil crisis. At the time, it was not a response to climate, but an attempt to respond to energy depletion or geopolitical instability. Japan launched the 'Sunshine Project' for renewable energy and the 'Moonlight Project' to improve efficiency. Japan has traditionally set energy security as a core national policy. There was a time in the mid-90s when research and development costs for Japan's energy technology stagnated. At the time, we were also developing a 10-year plan for energy technology development. I personally visited the Japanese Government and heard their reasons for the stagnation of the budget. At the time, the person in charge said that the Japanese Parliament always accepted the government's budget requests related to energy according to the original bill. He explained that this was because the performance of energy technology development was too low, and it was impossible to ask for more. It was an enviable and amazing experience from an energetic writer's point of view.

As climate response is in full swing now, the field of traditional energy technology is expanding. Technology groups that had previously been neglected, such as zero energy buildings, hydrogen energy, SMR, ESS, and hydrogen reduction steelmaking, have come to the forefront, and the pace of innovation has also begun to accelerate. This is because demand has increased due to government policy dissemination. Mass production effects were expected, innovation and convergence were promoted due to the influence of digitalization, and the intensity of innovation has also increased due to the tightening of climate-related regulations such as RE100. Policies began to lead the market and technology, but it wasn't until Bill Gates began preaching the importance of policies. Green racing, which existed only in words, began in earnest as competition among countries evolved, as each country raced to be the first to acquire technology that would provide a powerful solution to climate response.

Among these energy technologies, breakthrough technology is that which is difficult to innovate but guarantees significant effects when realized. In this regard, the '2050 Net Zero Forum', which was composed of more than 100 energy and climate experts to plan Korea's LEDS (Long-Term Low Carbon Development Strategy) plan, compiled key technologies for climate response. Proactively developing these technologies, commercializing them first, and securing a share in the global market is at the core of Green Racing.

<Long-term Low Greenhouse Gas Emission Development Strategies, LEDS>
The Long-Term Low-Carbon Development Strategy (LEDS) is a long-term plan established by the government on what changes need to be made to achieve carbon neutrality by 2050. Under the Paris Agreement, the parties must put in place a long-term plan on how to reduce their dependence on fossil fuels and submit it to the UN. To establish LEDS, the government formed the 2050 Low Carbon Society Vision Forum (Expert Forum) composed of 69 experts from various fields including academia, industry, and civil society. The Expert Forum proposed reduction technologies that should be implemented to achieve the goals based on the 2050 carbon neutrality scenario. Scenario 1 is the maximum emission target considering all reduction measures, but to achieve carbon neutrality by 2050, an additional reduction of 178.9 million tonnes compared to Plan 1 is required. Therefore, reduction measures to achieve further reductions were discovered, focusing on new technologies. The reduction technology

proposed by the expert forum is shown in the table below. It necessitates major changes relating to policy support, technology acquisition, technical stability, and reduction of cost burdens.

[Reduction technology that should be added to the LEDS maximum plan]

Carbon neutrality plan (forecast)	**Recommended maximum**	**Reduction plan**
Transitions	· 60% of renewable energy generation · 4.4% of coal power generation · CCUS collected 38.8 million tonnes	· 80% or more of renewable energy generation · 0% of coal power generation · Northeast Asia Super Grid
Industry	· 45% application of hydrogen-reduced steelmaking · Clinker recycling rate 50% · Smart factory 100% · Energy source unit improvement of 2% per year · CCUS 5 to 6% applied	· 100% hydrogen-reduced steel · Clinker recycling rate of 80% or more · Commercialization of innovative materials (biological raw materials, etc.) · Additional expansion of CCUS · DAC introduction · Zero allocation for companies of a certain size or larger

Carbon neutrality plan (forecast)	Recommended maximum	Reduction plan
Edifice	· Energy savings rate 24% (compared to BAU) · Distribution of high-efficiency devices 6% (compared to BAU) · Dissemination of HEMS (66%) and AMI (100%)	· Green hydrogen fuel cells replace city gas · Introduction of artificial intelligence buildings · Optimal design based on LCA of buildings · Mandatory zero emissions for buildings of a certain size or larger
Transportation	· Eco-friendly vehicles 93% · Autonomous driving and the spread of artificial intelligence · Advancement of shipping/railway/aviation	· Accelerating eco-friendly vehicles · Subsonic capsule trains · Fully autonomous vehicles
Government/Common		· Additional CCUS and DAC need to be introduced for those that have not achieved net zero by sector

* CCUS: Carbon Capture, Utilization, and Storage; HEMS: Home Energy Management Systems; AMI: Advanced Metering Infrastructure; DAC: Direct Air Capture; LCA: Life-Cycle Assessment

[Ministry of Environment, 2050 Long-Term Low-Carbon Development Strategy (LEDS), 2020]

Source — http://me.go.kr/home/web/policy_data/read.do?pagerOffset=0&maxPageItems=10&maxIndexPages=10&searchKey=&searchValue=&menuId=10262&orgCd=&condition.code=A3&condition.deleteYn=N&seq=7633

Green racing is not just an academic competition for scientific and technological achievements. The completeness of technology must first be proven through marketization on a certain scale. Cost-effectiveness must also be guaranteed by establishing a mass production system. Also, it is necessary to capture a certain share of the global market through unconditional overseas exports. Through this, the dominance of core patents and international standards must also be secured. Due to the nature of large-scale energy technology, it is difficult for one country to control everything. Therefore, a strong strategic supply chain must be built through solidarity and strategic collaboration between countries or companies. This series of processes is already underway. Let's look at the situation for each major technology.

For example, it is difficult for us to surpass China's global market share of solar power. However, Korea is second to China in this competition. Therefore, it is desirable to secure and maintain leadership in a specific market with technology that differentiates us from China. In the case of offshore wind power, it is difficult to secure a technological advantage in turbines, etc., but other components have characteristics that guarantee global competitiveness. Therefore, it is desirable to pursue leadership in manufacturing in the Asian market through strategic cooperation with Europe. A strategy of sharing a certain scale of market for this purpose is valid. Although it has low added value compared to LNG ships, the creation of jobs is expected to be quite large. In addition to the spread of renewable energy, industrialization is also an important virtue. The public sector did its best for industrialization during the Moon Jae-in administration.

In the case of nuclear energy, it is true that it received tremendous support from the Yoon Seok-yeol government. We are doing our best to export nuclear power plants. Nuclear energy also has various trends in terms of technology. In addition to nuclear power plants, various technologies such as nuclear fusion, pyro processing, and SMR are competing internally. Among them, securing a disposal site for spent nuclear fuel, which is a particularly long-standing challenge, is key to the development of the nuclear ecosystem. The safety of nuclear energy is an important issue. In this respect, apart from the issues surrounding nuclear power plant exports, the continued operation of nuclear power plants in the immediate term remains complicated as far as safety is concerned. As such, nuclear power plants themselves are complicated. In addition to support, regulation and surveillance are unavoidable. Collaboration with the United States, which is inevitable, is also an important issue. From this perspective, the role of the government in nuclear energy is greater than that of scientists. The way nuclear energy innovates is very different from other technologies.

In the case of hydrogen energy, there are no leading countries or companies in the world yet, so a strategy to preempt technology and market leadership by realizing a full-cycle value chain in Korea as quickly as possible is valid. Significant demand is expected in the mobility and defence sectors, such as hydrogen vehicles and hydrogen drones, and it is possible to create a market for fuel cells suited to decentralization even in the case of power systems. This large-scale strategic subsidy to generate demand is a welcome measure to secure hydrogen's global leadership. Of course, currently, due to KEPCO's large-scale deficit, the government is passive about the volume strategy of such acceleration, but considering the high density of hydrogen energy, it is a challenge worth taking on.

CCUS technology is also complicated. The feasibility of technology that captures carbon dioxide and buries it underground is heavily debated. However, extracting hydrogen from gas and disposing of carbon dioxide mass is the same as having to dispose of it underground after using nuclear energy. In other words, the core challenge of nuclear energy and blue hydrogen is the same in terms of the number of disposal sites. So, if nuclear energy is used, blue hydrogen must also be used. Blue hydrogen is an indispensable option to facilitate the use of fossil fuels as energy. Hydrogen energy is the best example of green racing, and it is an inevitable choice in terms of the energy mix in Korea.

In addition, there are so many other technologies that are currently being researched and developed. Most of these technologies are built into grids such as power grids. Efforts to systematically fuse these various technologies are also being strengthened. Who will be able to efficiently link to various infrastructures and grids, such as power grids, communication networks, and gas networks? This is the core value of innovation. Larger convergent technologies such as sector coupling have recently been actively discussed beyond IT-based smart grids. In other words, system improvements to eliminate partitions between energy sources are also inevitable. Technological innovation now requires more complex approaches such as product, platform, and system improvements beyond unit technology. As a result, there is a need for more public involvement and leadership. Green racing is huge and complicated.

The big question is which of the various technologies established earlier will be strategically positioned as a breakthrough technology. We have reached a point where we must solidify Korea's position in the global supply chain reorganization process. This is a time when the age-old topic of 'choice and focus' must be discussed. However, there's a huge difference between nurturing technology in a university research institute or national research institute laboratory and competing in the market. The technology must survive in the market and become a competitive winner, earning money, creating jobs, and paying taxes to the country. In the past, the government took the lead in deciding which technology to focus on, but now it is preferable for companies to make this choice. A medium to long-term roadmap for selected energy technologies shared by the public and private sectors must be secured. This is extremely important.

And we must act right away. Based on public-private consensus on a medium to long-term roadmap, Korea must achieve it first. To do this, the entire country must serve as a test bed for innovation. Also, an initial market must be provided through strong subsidies. Like Europe, we need a Korean version of the IRA law. In fact, Korea is already implementing significant policy measures stipulated in the IRA Act. However, collaboration between and within departments is lacking. Furthermore, Korea is making a mistake by prioritizing basic power supply and demand plans that determine where and who should build existing power plants. This is nothing more than the spread of traditional technology. A blueprint for breakthrough technology is more important than building a power plant.

The only harmony between the national interests of individual countries and the common response of humanity lies in technological innovation. This is the essence of green racing. Green racing is also competing with the speed of global climate change. Who will innovate faster? Prometheus, who created us humans and gave us fire, is watching things unfold with interest. The key to winning the green race lies in taking advantage of these breakthrough technologies first. I would like to reiterate this point to the members of Club Roma who wrote *Limits to Growth*. It's too late to turn our backs now. Even if the future is uncertain, things have reached a point where we have no choice but to rely on scientists and companies.

Chapter 3

Greening of Manufacturing Capacity and Energy Policies have Already Grown Together

Energy grew independent of the manufacturing industry, and as a result, it successfully played a role in supplying the fuel needed by the manufacturing industry. In terms of government organization, it started with the Ministry of Commerce and Industry and became independent as the Ministry of Power Resources in 1977 to foster energy systems such as electric power, oil, and gas. Although it is currently integrated into the Ministry of Trade, Industry and Energy, the reality is that it still operates separately in the form of an 'Industrial Office under the First Deputy Minister' and the 'Resource Office under the Deputy Minister.' However, these two brothers grew together by influencing each other. This reciprocity is the foundation for green racing, which must begin in earnest.

It is a mistake to perceive green racing as a new trend. The energy transition policy of the Korean Government in the past is just one type of fuel transition. There's nothing new under the sky. A good example of the success of green racing in Korea is the New Lighting Project promoted by the Ministry of Energy a long time ago.

In the early 90s, the government was actively engaged in energy-saving activities, and the popular tactic at that time to save energy was to replace the mechanical stabilizers of fluorescent lamps with electronic ones. The Ministry of Power Resources took a dramatic decision to change the fluorescent lighting systems (lamps, ballasts, luminaires) from 40W to 32W. The expectation was that if this market transition was successful, the overall electricity consumption would be reduced by about 4%. Towards this end, the government implemented comprehensive measures to improve efficiency in the lighting sector (New Lighting Project) by organically organizing and utilizing all means available to the government, such as large-scale research and development projects, KEPCO's subsidy system, minimum efficiency standard system, and cooperation with consumer groups. The Korean Government had a conversation with the lighting industry and subsequently started a market transformation programme based on partnership. This resulted in the fluorescent light that we currently see in our daily lives. In the past, fluorescent lamps were monochromatic lighting fixtures that were much fatter than they are now. Utilizing energy policy to drive market and industry innovation is indeed a prominent example of a best practice. This is the seed of green racing.

Through this strong, efficient regulation policy and a drastic market transformation, we have not only achieved rapid improvement in the efficiency of the lighting sector but have also strengthened the global competitiveness of white appliances.

At the time, there was a strong backlash from the industry against this attempt. In the case of a few consumer electronics, policy opposition from the Ministry of Commerce and Industry was considerable. Nevertheless, these two projects in the early 90s set the framework for developing efficient projects in Korea. Looking back, they seem to have been exemplary cases of green racing. It is essential to comprehensively promote the greening of the industry through the intervention of such strong energy policies for all sectors. Of course, the greening of the building and transportation sectors will also stimulate the greening of our manufacturing industry towards a more sustainable direction. It has become an unavoidable choice for us.

Green racing is also expanding into new business areas based on energy-related infrastructure. A typical example of this is the smart grid project that was promoted a long time ago. The power IT policy promoted by the Ministry of Industry during the Roh Moo-hyun government is a good example. In the early days, it was an upgraded project for energy, especially power systems, but when Jeju Island switched to a zero-carbon (CFI) project during the government of Lee Myung-bak, a new green business vertical of energy and non-energy was created. The convergence business, which began with electric power IT and led to CFI, still maintains its original form and operates as a model for green racing.

The universal characteristic of all these projects is the tremendous expectations and participation of Korea's leading manufacturers. It began as part of an energy policy, but I am confident of the emergence of a huge new industry. In addition to the various power systems upgradation projects, the focus was on other projects thriving at the time, such as DR, EV, ESS, and V2G. Eventually, the traditional manufacturing industry wanted to find a path to greening by utilizing traditional energy technology and infrastructure. The telecommunications industry is famous for propagating de-communication and trying to enter the energy sector in the name of Smart Grid. Green Racing aims to develop new markets and expand the added value through the greening of all industries.

This move of the telecommunications industry into new areas caused considerable friction with KEPCO at the time. KEPCO, which passively resisted the neoliberal government's competitive introduction and privatization policy, found it impossible to accept it because the telecommunications carriers perceived it as a real threat. In the end, this attempt at fusion failed to produce fruitful results. There were many experts who pointed out that this lack of support was due to KEPCO's excessive vested interests and monopoly orientation. However, contrary to the motivation of telecom operators, it is true that their lack of preparation and technological convergence also played a part. The creative business models were unable to exert their power due to the low electricity bills unique to Korea. This is because the low fees could only accommodate low levels of technology.

Since much of green racing is based on infrastructure in the energy sector, which includes electric power, collaboration and conflict with the traditional energy sector, especially the electric power sector, is inevitable in the future. In this respect, the open operations of the energy infrastructure and operating systems are inevitable. If energy changes, the industry may also change. This means that the tasks of the energy or power sector must be modified. Electric power must move away from its traditional task of supplying electric energy cheaply and stably to economic entities and focus on helping the existing manufacturing industry to achieve high-added value, respond to climate regulations, and secure global leadership. In a nutshell, its responsibilities must be further expanded.

Power must ensure that the hardware infrastructure is more open and robust, and the operating system must be modified to be more transparent and flexible under the expanded mission that guides green racing. And consumers of electricity must be willing to accept flexibility in electricity bills, or price increases so that the power sector can perform these new tasks. Innovation in hardware, software, and fee systems in the power sector is critical to ensuring the greening of our manufacturing industry. If we do not accept this, the greening of our manufacturing industry will gradually lose its competitiveness. Also, our wealth will be reduced, and future generations will lose their jobs. This is all too obvious. The source of power itself must change. Only then can our manufacturing industry survive and continue to evolve.

Chapter 4

Green Racing's Energy Policy is a New Industrial Policy.

We can't ignore the Paris Agreement, which is an international agreement. We depend on exports to eat and live. In the process of reorganizing the global supply chain, an industrial policy to preserve Korea's factories is critical. Only if this is incorporated into our energy policy can the country be freed from WTO and the like. The greening of Korea's manufacturing industry is inevitable, and artificial government intervention for this purpose is also inevitable. Government intervention means, in a very practical and political sense, the direct intervention of energy policy measures in industrial policy.

What is particularly important is that Korean companies competing globally want the government to take the lead in green racing. Our companies can only invest in green racing if our government takes the lead. As Bill Gates said, not only technology and markets but also policies are necessary to respond to climate change. However, our government's active market intervention and support are regressing, and there is a growing concern that private investment will also shrink due to this. But it is only under the condition of a collective effort between the government and private sector that this can be achieved, just like in the Battle of Jinju Castle.

It is difficult to judge the performance of such a large-scale government-led project. This is because all businesses have their own seeds of growth and evolution regardless of success or failure. However, the problem is that over time, more and more businesses are focusing on mere calculations of cost and profit. This is clearly the fault of the relevant department on paper, but on the other hand, the political intervention of elected officials is a bigger problem. It is a reminder that policy stability is so important for large-scale projects that require the approval and support of the Central and local governments, and the cooperation of residents.

As explained earlier, we have a lot of experience in promoting large-scale government-led projects related to energy. However, the results have not always been good. The Offshore Wind Farm Project promoted in the Seonam Sea during Lee Myung-bak's administration eventually became a tomb for large companies that promoted offshore wind power in Korea. Currently, the Floating Offshore Wind Power Project off the coast of Ulsan is also in jeopardy. Hydrogen projects promoted on a national level are facing challenges. The lessons that must be learned from watching this series of events unfold are the importance of policy consistency and political support to bear the cost burdens. When I think about the unwavering commitment of the government to the construction of the Gyeongbu Expressway during a time when there was no demand for logistics and a limited budget, I'm once again amazed. Now *that* is the true role of the government.

Given this situation, the new task assigned to the energy industry is not to support the industry from behind but to drag it forward. The energy industry itself, not the government, must take on this mission.

For example, consider KEPCO's current deficit problem. What kind of manufacturing industry do you want to help when the deficit in 2022 was 30 trillion won? Moving forward, we must solve this problem. Should we reduce the deficit to 29 trillion through painstaking efforts such as prohibiting employees from using corporate cards? Or should we induce private investment in related industries such as hydrogen, offshore wind power, and electric vehicles using 1 trillion to generate demand for innovation, even though it costs more money, and the deficit increases slightly? A decision must be made. Which of these two scenarios is in line with our national interests? Of course, it is right to induce investment for the future. Green racing is also ridiculous if the timing is not right.

Can the green industry, which is supported and led by energy, take control of global hegemony? It's possible. Using Korea as a huge global test bed can overcome the limitations of scarce technology and capital. However, first, the government must respect the choices and focus of our companies and promote system improvements that support them. But the most important thing is the speed of innovation! Fast energy transition and the rapid greening of the industry are important. It is possible to secure global leadership as soon as possible. Increasing the speed of transformation increases our chances of winning. This guarantees the demand for innovation through accelerated subsidy support for each technology. This means that technological innovation must be promoted quickly. It's literally a green race. If you invest quickly and innovate quickly, you will become a strong player in the global market. Too many companies around the world are already participating in this race. However, the current investment and regulatory conditions in Korea are too messed up.

The government must integrate energy, industrial and climate policies. It must also establish policies to maximize the motivation of companies to invest. Decision-making methods used over the past 60 years must be reinvented. Maintaining the consistency of these policies is paramount. If the system and direction are changed every time the regime changes, our companies must move their factories overseas. We also need to seriously consider how to get support from citizens and consumers for integrated policies. The first thing I did when publicizing spent nuclear fuel the other day was to make people aware of the spent nuclear fuel issue. As more people understand the issue, our work will be easier.

What is needed right now is an upgradation of the regulatory system that will allow companies to invest. Currently, most of the government's investment resources are covered by energy bills. In other words, the money that energy consumers pay is capital. It uses money paid by energy consumers to build, operate, and provide public funds for the future. However, this process is operating under a very complex trading system called the 'power market'. The government's policy is mechanically operated in the same way as a 'ladder climbing' game where different players strategically position themselves. Therefore, investment in innovative fields is induced only by delicately and precisely designing market rules that are responsible for investments and transactions independent of the consumer's intention to pay. The sum of these rules can be said to be a package of regulations. Advancing these regulatory packages is key. If you only intervene in the market with goodwill, it will eventually lead to the same results as the government's real estate policy in the past. Market liberalization and competition alone are difficult. The advancement of regulations is literally a virtue that requires a high level of expertise and meticulous courage.

The most serious obstacle to innovative investment is location. In addition to the urgent disposal sites for spent nuclear fuel, new locations such as offshore wind power and hydrogen stations are essential. This poses a huge challenge for us. Therefore, the use of existing locations is inevitable and urgent. First, there is an urgent need to increase the transparency of regulations for investment and improve systems to facilitate one-stop licensing and planned locations. The realization of electricity bills is essential. To generate innovative domestic demand to induce investment, drastic technology regulation is also necessary. Securing a position is a factor in the new industrial policy. The five-year economic development plan that led to Korea's industrial development was essentially a location plan. We must make a fresh start now. The need of the hour is a new industrial policy.

Part 4: Three Enemies Preventing Green Racing

In this green racing, each country fights for the victory of its own industry over national interests under the guise of climate change response. As far as green racing is concerned, Korea started relatively early, but it seems to be falling behind. I was uneasy about Moon Jae-in's government, which was nothing but slogans, while Yoon Seok-yeol's government lacksdirection. I reiterate that the future of Korea depends on whether it can maintain its current status as a manufacturing powerhouse. Semiconductors and nuclear energy alone are not enough. Korea has more technological opportunities. Depending on the amount of wealth and the number of jobs, the greening of the industry can be expanded to become more significant. Green racing is essential to consolidate our status as a manufacturing powerhouse.

However, there are many factors that hinder and threaten this race. Among them, there are three most serious enemies. It is important for our consumers and citizens to know about these three enemies. These enemies don't exist on a serious level in our competitors, but they are internal enemies that are too difficult for our energy and manufacturing companies to battle.

Ironically, our past achievements will prove to be a disadvantage in the green race, with our narrow worldview and excessive selfishness.

As these achievements become more and more powerful, our chances of losing the green race gradually increase. These three enemies will ruin the future of our descendants. It cannot be left to elected officials. The only thing that can defeat this serious enemy is the renewed commitment of consumers, citizens, and specialized bureaucrats. Consumers must be willing to support the increase in energy charges to bear the increased costs of green racing. Voters must continue to support Korea's traditional energy diversification policy. Public officials must support the green racing of manufacturing companies more publicly than other countries. Only then can we win the green race.

There are other groups that will support green racing to strengthen our position as a manufacturing powerhouse. I look forward to their support also.

Chapter 1

Populists Who Insist on Cheap Energy are Enemies!

We are still paying huge sums to operate the world's highest-density energy. Energy accounts for 1/3 to 1/4 of Korea's income. Furthermore, additional investment is needed for climate response and green racing in the future. So how are we going to raise this money? The current method of financing is to utilize the money generated from consumers paying their energy bills.

<Korea's Electricity Bill Types and Structure>
Korea's electricity bill system is divided into residential, educational, industrial, agricultural, street lighting, and general use. In addition, progressive and differential pricing plans are implemented depending on usage, season, etc., so there are differences depending on the tax classification. However, the basic structure consists of basic charges, electricity charges, climate environment charges, fuel cost adjustment fees, VAT, and power industry infrastructure funds. For example, if an ordinary household uses 300 kwh of residential (low voltage) electricity during the month of January 23, the bill is 55,546 as per the configuration shown in the figure**.

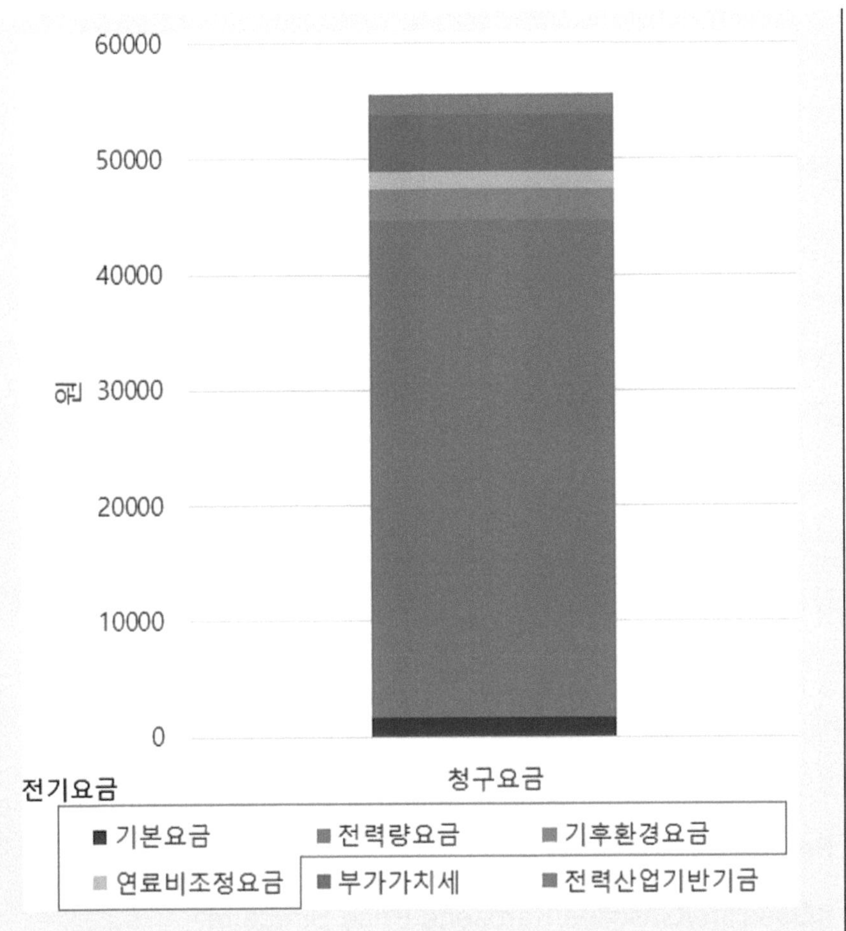

* Residential (low pressure) 300 kWh usage standard for January 23

Next, if we look at the price structure of major petroleum products, in the case of gasoline and diesel, it consists of the refiner's price, transportation energy, environment tax, education tax, driving tax, sales levy, and levy tax. In the case of kerosene, it consists of the refiner's price, individual consumption tax, education tax, and levy tax.

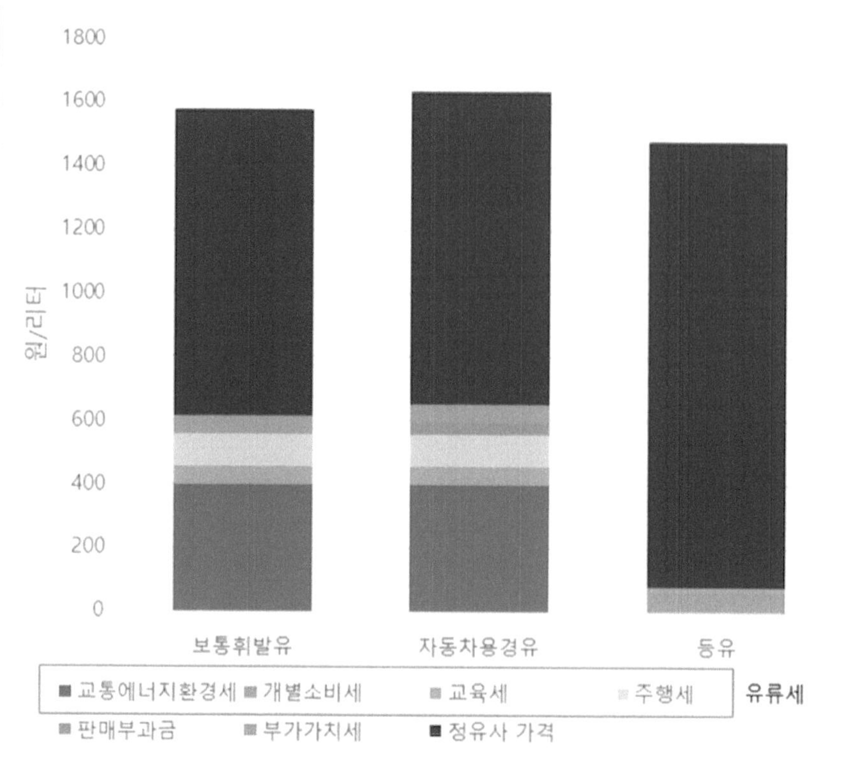

* Based on the national average for the second week of February 2023

<Gasoline/Gasoline/Kerosene Price Structure>

[KEPCO Cyber Branch, Electricity Bill System by Application]

[KEPCO Cyber Branch, fare calculation comparison, 2023.02.16. search]

[Petroleum Corporation OfficeNet, Domestic Oil Price Trends for the 2nd Week of February 2023, 2023.02.]

Source-https://cyber.kepco.co.kr/ckepco/front/jsp/CY/H/C/CYHCHP00201.jsp

https://www.opinet.co.kr/user/oftvat/getOftvatSelect.do

Aren't our consumers frustrated about paying this fee? A few people may be paying unnecessary fees due to inefficiency. Or maybe they're paying unwarranted fees due to someone's greed or carelessness. You might be curious to find out. Having worked in the field of energy for a long time, I can speak with confidence. Korea's energy services are excellent both quantitatively and qualitatively. At least for now.

Korea is an old story, but there was a time when power system innovation was promoted around the world under the name of 'smart grid' during Lee Myung-bak's administration. A senior official from the US Smart Grid Association, who led this effort at the time, visited KEPCO and made a somewhat arrogant announcement. At the time, when a KEPCO practitioner informed him about the power outage time and transmission loss rate in Korea, the official immediately assumed a humble attitude and expressed respect. As such, our power system was very efficient, and the fees were also very low. This is the case even now. The price of oil became an issue during Lee Myung-bak's administration. There were so many suspicions and complaints about oil companies that the President even directly called it 'strange.' As a result, a privately oriented price monitoring system called the Petroleum Market Watch Group was launched. There is probably no case where specific products are monitored so thoroughly and continuously. The Group concluded that there was no major problem with the price of energy in Korea, despite the suspicions.

<International Energy Prices>

International crude oil prices change to reflect the international situation. There were two oil shocks in 1973 and 1979, and crude oil prices repeatedly rose and fell due to large and small global issues such as wars and financial crises.

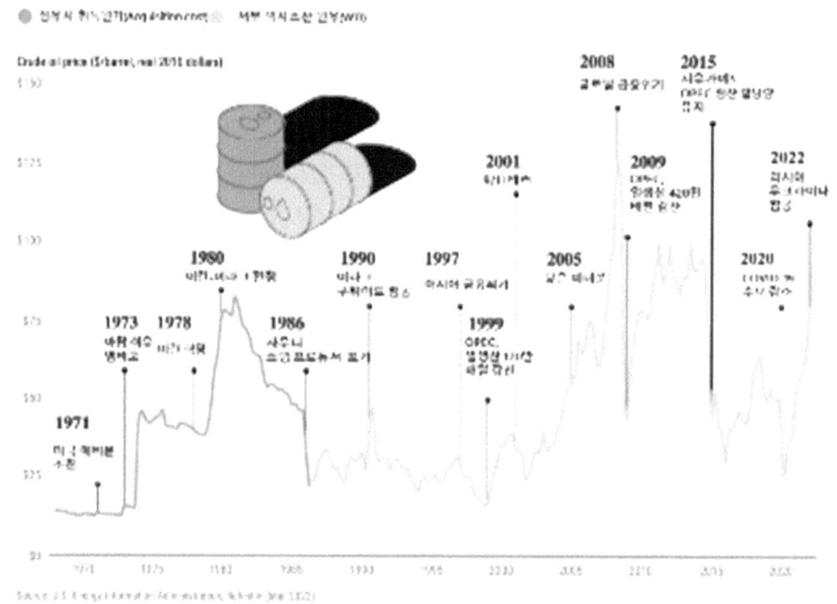

<Changes of Oil Prices Due to Global Issues>

Korea imports crude oil into the country through long-term contracts of at least one year with the governments of oil-producing countries and state-owned oil companies. For needs other than this, in-kind contracts are used to purchase a specific quantity of crude oil at a specific point in time. Therefore, crude oil is stably supplied domestically at a level similar to that of other countries, and the price of domestic supply is based on unleaded high-grade gasoline which is also stably supplied at a level similar to that of other countries.

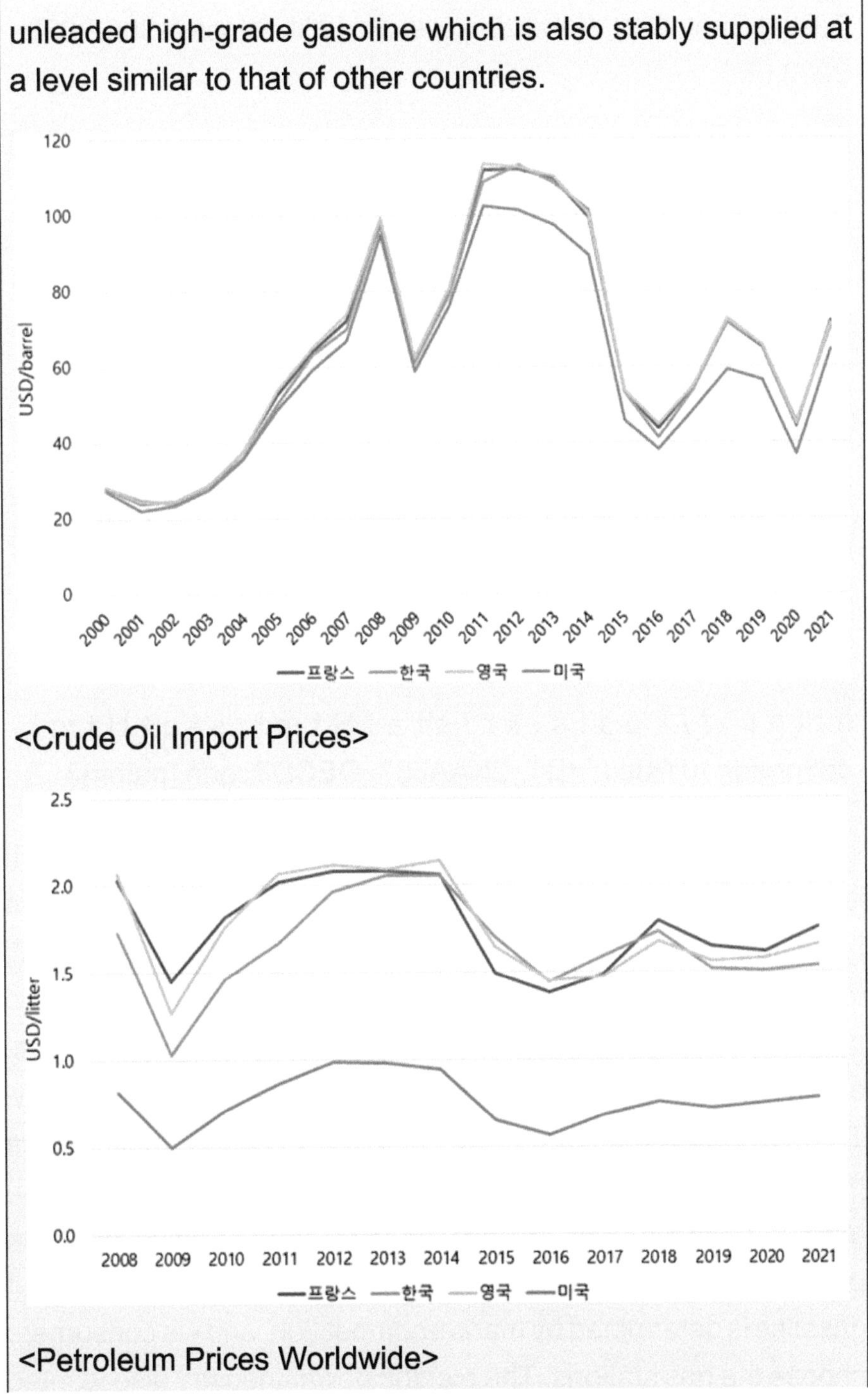

<Crude Oil Import Prices>

<Petroleum Prices Worldwide>

[Visual Capitalist, Visualizing Historical Oil Prices (1968-2022), 2022.05.]

[Korea Petroleum Corporation, domestic crude oil import type]

[OECD, Crude Oil Imports, 2023.02.12. search]

[Bureau of Statistics, Energy Prices (OECD), 2023.02.11. search]

Source — https://advisor.visualcapitalist.com/historical-oil-prices/

https://www.knoc.co.kr/sub05/sub05_9_1.jsp?page=2&field=&text=&mode=list&bid=CUSTFAQ&ses=USERSESSION&psize=12

https://data.oecd.org/energy/crude-oil-import-prices.htm

https://kosis.kr/statHtml/statHtml.do?orgId=101&tblId=DT_2KAA607_OECD&conn_path=I3

Who determines these energy prices or electricity bills? It is largely government-led or market-driven. In this regard, we have two experiences – petroleum and electricity. In 1997, the government implemented liberalization measures for the petroleum industry. Price liberalization, entry liberalization, and import/export liberalization were handled simultaneously. Through these measures, the petroleum sector, which was a public sector, transformed to a private (SK, GS, S-oil, Hyundai) market. Consequently, we experience ever-changing gasoline prices in our daily lives. There are still suspicions of oligopoly, but gasoline is determined by market competition, and we consumers choose the gas stations. The electric power industry also wanted to implement liberalization measures. Partial liberalization made private sector entry successful, but measures to implement price liberalization and sales openings were not successful. Success depends on whether the government can persuade consumers and businesses to buy into its policies. Oil was successful, but electricity failed.

President Kim Dae-jung's attempts to liberalize the electric power industry, led by a group of experts advocating marketism, did not produce the expected results. There was no mention of price liberalization, and the driving force was undermined due to hasty propagation that restructuring would reduce electricity bills. I missed the opportunity to adequately explain the concept of price liberalization to consumers, where electricity prices fluctuate in real time, particularly with varying rates in Seoul, Busan, and Gwangju. The communication just lacked clarity. There was probably a prejudice that prices would be automatically liberalized if KEPCO, which was popular at the time, was dismantled. In a nutshell, it wouldn't be wrong to say that the expert group that supported this understood the market and consumers even less than the bureaucrats who pursued oil liberalization. When the petroleum industry was liberalized, bureaucrats prepared consumers for price changes in advance before the liberalization measures were implemented, by displaying prices based on the exchange rate at gas stations. As a result, electricity bills are still determined by the market, even though they are set by the government. A typical example of this is the introduction of the SMP upper limit system.

The three most important taxes of this method to set the fees are represented by the perspective of the energy system, of macro-price management, and of industrial support. Of course, it is of significance for the government department responsible for prices to try and suppress the increase in electricity bills, which has considerable influence and is highly symbolic. Departments that implement industrial policies have no choice but to be passive about fee increases. Therefore, adhering only to the stability of an energy system can be evaluated from a selfish perspective in a larger framework. Finding balance through conflict and communication between these individual policies can be a legitimate process of policy formulation by the Korean government over a long period of time. However, it is understood that in the past, when KEPCO's deficit reached several trillion won, the government allowed and managed rate increases. Over time, the influence of political judgement in the political sphere began to increase. In other words, the practice of one-sided fee control was strengthened. Meanwhile, the demand of KEPCO's management team, that the interests of KEPCO shareholders listed on the New York Stock Exchange should also be guaranteed in response to this, gradually began to intensify. As a result, significant progress was made in the fee-rate system called the fuel cost indexation system. This system didn't really work, and then the crisis in Ukraine broke out. Flexibility in fees is just as difficult.

Is this kind of government fee control desirable for the national economy? Is it good for the electric power industry or for individuals involved in the electric power industry? As with any complex decision, there are pros and cons. The relationship between the government and the market has entered a strange era. The current situation is unfavourable, but neoliberalism is inappropriate, and a return to development dictatorship is not the right answer. This is not an age where one party can take the lead. A new delicate balance must be found. Green racing is a race where you can run at high speed on a narrow and dangerous mountain path rather than drive on a wide expressway. The navigation system that drives this race is the price signal. Charges, or prices, work magic. In green racing, people invest by running and looking at the barometer of electricity bills. Among these magic tricks, the most important are externality effects and grid parity. Let's take a closer look at what this magic trick is.

On one occasion, an economist claimed that externality was just a theory that appeared in textbooks. However, our government has been aware of the cost, or externality issue, that cannot be solved in this market for a long time and has raised and operated public funds to resolve this problem. Its history is quite long. The most typical externality-based charge is fuel tax. Also, the Petroleum Project Fund (now the Special Energy Account) and the Electric Power Industry Infrastructure Fund are the most typical devices for external processing. There is debate about the appropriateness of this scale and the efficiency of its use, and continuous efforts are being made on how to improve it. Our society will find the right answers.

<Externality Theory>
Externality refers to a market participant's economic behaviour that has an unintended influence on a third party or society, but compensation is not made for it. When it has a positive impact, it is called a positive externality or external economy; when it has a negative impact, it is called a negative externality

or external uneconomy. For example, when smoke is generated due to factory operations, people around the factory are harmed, but there is no compensation for this. This is a negative externality. On the other hand, if a beautiful flower bed planted on my compound wall gives aesthetic pleasure to people passing by, I get no compensation for this. This is a positive externality. Because of externalities, the market doesn't work perfectly. Negative externalities can be eliminated by government intervention, internalizing appropriate costs, and imposing taxes. Similarly, positive externalities can be internalized by compensating the unintended benefits through subsidies or the like. [Centre for Economic Information, Externalities and Market Failure]
https://eiec.kdi.re.kr/material/conceptList.do?depth01=00002000010000100008&idx=134

<Grid Parity>
Grid parity means that the cost of electricity production through renewable energy such as solar power or wind power is equal to or lower than the cost of electricity production through traditional energy resources such as coal, natural gas, and nuclear power. This means that renewable energy sources can

compete with traditional energy sources without government subsidies or incentives.

To compare electricity production costs based on the energy source, the Levelized Cost of Power Generation (LCOE) of power plants is used. The elements that makeup LCOE include initial investment costs, maintenance and operation costs, interest costs, power generation, generator life, and discount rates. It is achieved by increasing the LCOE of traditional energy sources due to the declining LCOE of renewable energy thanks to technology development, internalization of external costs (environmental costs, etc.), and rising fuel prices.

Since grid parity is affected by installation costs, efficiency, and support policies of renewable energy, it is used to analyze relevant business conditions and prospects of each country. According to the International Energy Agency's 2025 LCOE forecast, the median LCOE value for utility solar power in the US is 43.1 dollars, which is lower than nuclear power and coal power generation (71.3 dollars and 88.1 dollars, respectively), so grid parity is likely to be achieved. However, in the case of Korea, the LCOE of nuclear power and coal power generation is expected to be 53.3 dollars and 75.6 dollars, respectively, and solar power for utilities is expected to be 96.6 dollars, so the possibility of achieving grid parity is low.

[Korea Institute of Economics and Management, (16/24) KEMRI Power Economy Review, Grid Parity (Grid Parity)

Achievement Factors and Characteristics Comparison in Major Countries, 2016.10.] [Korea Energy Corporation, Grid Parity for Renewable Energy to Be Cheaper Than Fossil Fuels, 2021.11.] [International Energy Agency, Incentives of Renewables 2020, 2020.12.]
sources–https://home.kepco.co.kr/kepco/KR/ntcob/ntcobView.do?pageIndex=12&boardSeq=21026367&boardCd=BRD_000271&menuCd=FN3120&parnScrpSeq=0&searchCondition=total&searchKeyword = http://blog.energy.or.kr/?p=24859 https://www.iea.org/reports/projected-costs-of-generating-electricity-2020

However, fees in Korea are structurally cheaper than in other countries. Since we generate a lot of electricity, transport it in a short time, and consume it in large quantities, our fees are lower than in other countries. It is also because of the political preference for lower-rate policies. In the past, when I was in the Ministry of Energy and Resources, government officials said that Korea always set lower fees than their competitors. Thus, under the current fee structure, it becomes difficult to achieve grid parity in terms of accepting new technologies. Low retail prices have the advantage of supplying cheap energy to consumers, but on the other hand, they prevent the entry of 'good and innovative technology,' which is more expensive than the fees paid for energy consumption.

In other words, grid parity acts as a reference point for the market to accept innovative technologies required for climate response or green racing. If fees are low, the burden on consumers will be small, but it will be difficult to innovate, and the evolution of the industry will be suppressed, maintaining its traditional status quo. On the other hand, if the fee is high, the burden on consumers will increase, but green racing will be facilitated. Recently, I heard the opinions of an employee of a Danish Energy Agency. Denmark is the most exemplary example of an energy transition. "The energy transition was carried out well by raising the fees, and the related energy industry was fostered spontaneously by this." Indeed, (retail) charges are a magic wand.

If the government continues to control the cost of electricity and prefers to maintain the status quo of low rates, as is the case in Korea, it will remain in its comfort zone. The impact of the recent crisis in Ukraine, a tenfold increase in European rates, was shocking. However, we are responding to the sharp rise in fuel costs with a deficit that runs into trillions for KEPCO through charge control. Our government wants to manage KEPCO such that it doesn't go bankrupt, with minimal fee increases. Even though the company is on the brink of bankruptcy due to a huge deficit caused by an unreasonable fee containment policy, the KEPCO Union is not raising issues. Although there is some suffering, KEPCO's deficit is probably because it also guarantees its exclusive status. The government that wants to seize prices, KEPCO unions that prefer exclusivity, and consumers who pay less will all be happy with this artificially low-rate policy. However, these policies will not be fruitful in the long run, electricity inefficiency will increase, its quality will deteriorate, and fine dust will increase. The advancement of energy systems will also be curbed, and our manufacturing industry will remain a low value-added industry. Green racing in this situation will be impossible. More importantly, KEPCO's deficit is just external money that consumers must eventually pay because they will end up paying interest later.

However, the biggest enemy of rate normalization is the political sphere. The political belief that the increase in electricity bills will lose votes is too strong. Currently, the entire government is criticizing the delay in rate increases, but there is still much to be done about the intensity of the rate increase. Europe has a system that allows a tenfold increase in retail prices. It is true that we are still lacking compared to Europe. Of course, transforming KEPCO's 30 trillion deficit into a 20 trillion deficit is insufficient. In a situation where the cumulative deficit is expected to reach 50 trillion, there is still not enough real persuasion from the public about fees.

The belief that these price increases will eat away at the ticket is like a ghost. There are no official claims anywhere, and there are no official documents. Perhaps because he was afraid of the ghost, Minister Mo, who led the deflation war during the Moon Jae-in administration, began to distort the energy transition policy by saying that it would not cause price increases. I should have proudly explained the costs of the energy transition and climate response. I think many people are suffering because of this cowardice right now. Ghosts have no substance, but they exist and prevent healthy debate. The ghost becomes even stronger ahead of the election. The government of Lee Myung-bak also insisted on green growth but avoided mentioning the cost, or energy bill, which is the pain associated with it. However, it failed to recognize the maturity of our citizens who have achieved democratization and industrialization. I am confident that if Korean consumers are provided with sufficient explanations and relevant information, they will sympathize and accept an increase in electricity bills.

We should also reconsider whether there would be only opposition to tariff increases. During the Park Geun-hye administration, a related industry discussion was held at Daehan Sang regarding the increase in electricity bills. Companies were largely composed of energy-intensive industries such as cement, energy industries such as power generation companies, and high-value-added industries such as semiconductors. These three groups have different positions on electricity bills. Therefore, an official from Daehan Sangsang said that it was difficult to formally express a unanimous opinion on the electricity bill. The interest in the electricity bill is more complicated than one might think. When the electricity bill per kwh rises, the burden increases, but in the end, changes in related technology and behaviour also reduce the cost for electricity consumers. Again, the old saying, “Always cheap is *biji-mochi*” is wise.

The existing cheap energy is white rice, and the more expensive energy of the future is organic brown rice, which is good for health. Aren’t we taking all kinds of expensive nutritional supplements for our health nowadays? Cheap energy, or white rice, is an unhealthy food that destroys our bodies and our manufacturing industry. Even if it’s a bit expensive, we are now close to the point where we should eat expensive organic food. Prices in Europe have increased tenfold this time, and the ecosystem has become a muddy field. On the other hand, we are likely to be relieved that KEPCO's deficit blocked the agitation of the rest of the economy and stopped inflation. However, in the future, Europe will expand investment in this field and the ecosystem will evolve, and we will stick to our current methods and fall behind the times. Of course, it is obvious that private investment will shrink; it will collapse due to the next external shock. This is what populism is all about.

However, if the Korean people explain it well, energy consumers will be willing to pay. After all, if you pay money, a lot of good things will happen. The cheap energy is like *biji-mochi*. It makes sense for the price balance point to be adjusted through policies along with the market. We must provide food to our manufacturing industry so that it can lead the global market. We must generate demand for innovation in advance so that our companies can get a head start in the global market. Paying costs is just an investment for raising twins to become stronger adults. Don't skimp on energy costs. We must pay more than we do now. And if we pay whatever money we need to pay as soon as possible, our twins will grow healthier. That's it.

This is because we have now reached a level where we can eat organic energy. The feasibility and cost-effectiveness of paying for energy, supply and demand, and greening should be analyzed in detail and discussed and explained transparently and openly. It is possible to create sufficient consensus based on the conflicting responses of the various stakeholders. We can gain the understanding and consensus of the public through transparent disclosure of information and passionate discussions. This requires the professionals to play a courageous role. Failing to raise the fees now will cause problems in the future. It's all an expense that consumers will pay for anyway. Politicians who have secretly been traumatized should never be applauded. It's just a hoax.

Chapter 2

'Monocolorists' Obsessed with Specific Energy Sources are Also Enemies!

The foundation of Korea's energy policy for the past 60 years was absolute fuel diversification, and success has been achieved through it. After 1962, energy was supplied to intensively support economic development plans. In 1962, half of Korea's energy was fuelwood. However, after that, they began to diversify into briquettes, gas, and nuclear energy. In particular, the oil crisis around 1980 was an opportunity to further strengthen this diversification policy. At that time, about 70% of power plants were petrol-fired, but the Republic of Korea quickly put an end to petrol-fired power plants.

There are many stories about policy trends related to energy supply and demand. Beginning in the 1990s, officials from the Ministry of Power Resources decided that it was difficult to maintain an infinite supply priority, and they tried to shift to a demand management centre. I've already begun to experience trouble myself. However, it was defeated by KEPCO's backlash. Also, there's a call for changes in the overall cost principles due to the reorganization of the electric power industry. Price reduction was anticipated by introducing competition and market functions. This movement of change, however, ended with a formal change. As the years passed by, the Economic Power Supply Principle under the Electricity Business Act, which had been firmly followed for so many years, was revised to 'Economic Charging That Takes the Environment into Consideration' by agreement. However, nothing has changed since then. Korea's electric power sector has not deviated from the 1962 paradigm.

Since then, with the advent of fine dust, climate change NDC, etc., environmental issues have begun to change, bringing focus to the electric power industry. In particular, the power sector began to experience rapid changes as environmental groups targeted the transition sector (electric energy) over the industry, seeking reductions. A real change in emphasis on the environment began when consensus was created on drastic reductions in coal power generation. President Moon Jae-in's declaration of carbon neutrality in 2020 was a major event, highlighting the environmentally focused policy. However, at that point, the Moon Jae-in government's anti-nuclear war policy clashed with the carbon neutrality priority policy, causing a full-scale conflict. As energy policies became the target of full-scale strife, the energy sector experienced structural conflict and a decline in capacity. Korea's energy is essentially stuck in the 1960s.

Also, the war in Ukraine, which began in 2022, was an opportunity to announce once again the strength of the existing fuel systems amidst climate-centred thinking, or carbon neutrality. The gas market began to fluctuate, creating global inflation, and gradually causing enormous conflicts such as supply chain confusion and capital market confusion, which was another opportunity to confirm the strategic importance of energy. It was also an opportunity to acknowledge the correlation between the status of a manufacturing powerhouse and energy security, through the case of Germany. The US IRA law has once again become an issue that we must look closely at.

Here are some of the most memorable and disappointing scenes. The transition to a demand-management-centred energy policy implemented by the Ministry of Power Resources, the setback of an entire academic research project to fundamentally improve the power grid during the government of Lee Myung-bak, the transition from the Park Geun-hye government's Third Basic Energy Plan to a clear decentralization strategy, and the transition to a hydrogen society during the Moon Jae-in government all ended in failure. If we had adopted such measures in a timely manner, I'm convinced that we would have been able to become the strongest manufacturing nation through green racing. I wonder what some officials in the Ministry of Industry and some employees of the Energy Corporation, who opposed these policy changes at the time, feel about the situation now.

The conventional standard of affordability usually lies at the root of this regret. Large-scale energy transition policies are bound to involve enormous costs. It is true that it is difficult to rationalize the costs of climate response and greening-related industries with normal economic efficiency. However, if we use the economic efficiency measures currently used, we wouldn't have even tried to build the Gyeongbu Expressway. Also, fuel transitions such as the introduction of nuclear power or gas would not have occurred. Affordability is simply a reference point.

The other day, I visited the Japan Institute of Energy Economics and interviewed Japan's basic energy planning team. The experts asserted that economic efficiency is only one factor that determines the energy mix. The power generation facilities put in place in Korea during the Jeon Doo Hwan administration were extremely excessive, and it was clearly a policy failure if you look at it as a measure of economic efficiency. However, the three lows occurred at the time, and our economy took a big leap forward due to its high reserve power.

<Energy Mix Optimization Price>
Recently, the Energy Mix Special Committee (hereafter referred to as the Special Committee), which includes the Nuclear Society and Seoul National University, analyzed changes in power generation costs and the impact on electricity bills due to various energy mix scenarios. Energy mix refers to the procurement of energy demand by combining various energy sources. Energy mix optimization is structured considering national plans and the economic efficiency of power plants. The Special Committee organized the 2050 energy mix into four scenarios, considering the nuclear power plant policies, when to close coal power plants and the share of renewable energy. According to the analysis of the results for each scenario, it was found that Scenario 2, with the highest share of renewable energy, had the highest cost of power generation. However, among scenarios with an equal share of renewable energy (Scenarios 1, 3, and 4), it was observed that Scenario 1, which had the highest share of gas, had the highest generation cost. In the future, the cost of power generation for each energy

source may change as grid parity is achieved and social costs are internalized due to externalities.

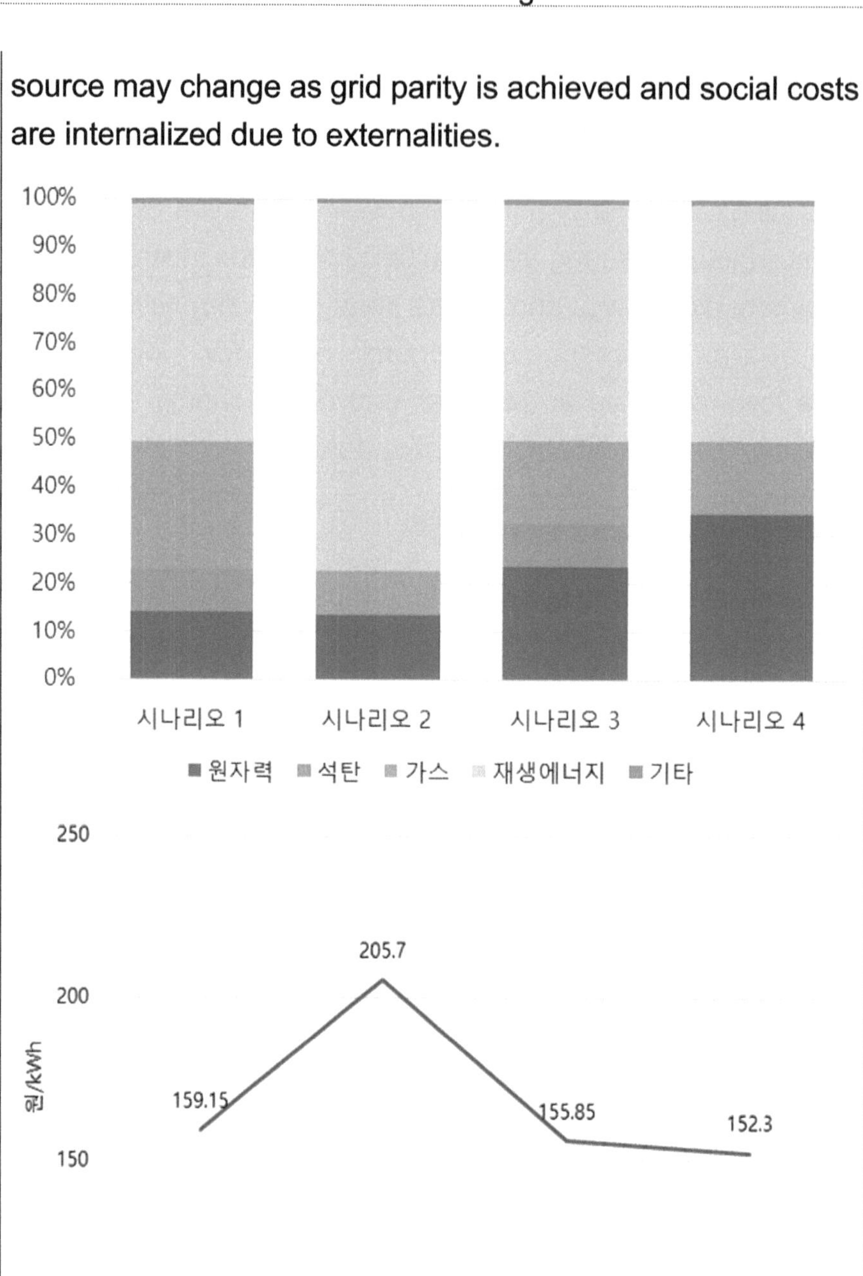

[Energy Mix Special Committee, Energy Mix Analysis Report, 2021.08.]
Source - https://www.kns.org/boards/view/notice/101028/page/

It makes perfect sense that we transform our infrastructure right now. Our reliance on coal must be reduced and nuclear power plants, renewable energy, and hydrogen energy must be expanded. To do this a huge investment is necessary, which means we must increase the fees. This point must be clearly explained to consumers so that they are convinced. This is what is required from the current leadership. Green racing can only be achieved by producing energy efficiently and innovatively, and this is what drives the interests of consumers.

What kind of fuel should we use to support our society in the future? It is right to convince consumers and citizens by clarifying this point. Of course, we must consider the conditions we have no choice but to accept. NDC, RE100, etc. must be accepted. Also, in situations where telephones are unavoidable, decentralization that considers communication and system problems must also be accepted. To maintain the status of a strong manufacturing country, nuclear power and renewable energy alone are impossible to meet the supply and demand of energy, so technical measures to utilize existing fossil fuels are necessary, and the implementation of a hydrogen society using CCUS, and the like, is also inevitable.

All these options are not possible based on the cost-equilibrium economic criteria provided by existing energy systems. It is necessary to pursue transformation based on innovative technology as the driving force behind these tremendous changes. With the advent of hydrogen reduction steelmaking methods, electric vehicles, hydrogen vehicles, and zero-energy buildings must be pursued simultaneously, which will increase costs. We must utilize the wisdom gained from building the Gyeongbu Expressway, where a huge sum of money was invested without demand. Again, the price balance point must shift.

What is particularly important here is the strategy for the process. Only by implementing this transition faster than other countries by taking advantage of our experience in building the Gyeongbu Expressway, can we secure technology and manufacturing capabilities that guarantee global competitiveness. This will help us win the green race. In resource-poor countries such as South Korea, this innovative approach to technology is a crisis, a challenge, and the only way to live. Consumers and citizens must join in this effort to save our manufacturing industry and our jobs.

In this way, we must make a quick transition. However, as always, this evolution is fraught with difficulties. There are groups that decisively interfere. They reject diversification and impose only specific goals. They are 'monocolorists'. These 'monocolorists' are groups that oppose Korea's energy strategy. Any group that favours or hates special powers (some political circles and civic groups based on rigid environmental ideology, and groups that only stick to the interests of special agents) is the enemy. They are politicized groups that have fallen prey to distorted petty heroism, not based on national interests and intellect, and which secures their position through strife.

They seek to secure private power through such an intolerant worldview.

These monocolorists shook policies with each regime change, expanding investment uncertainty, and consequently reducing the willingness of companies to invest. In the new NDC reset process, an appropriate mix must be set through sufficient discussions on the three axes of carbon neutrality, energy supply and demand stability, and the greening of manufacturing. This is something that must be solved within the institutional sphere (especially the governing body, Tan Jung-yu, etc.). If you look at recent demineralization debates, solar debates, and Saemangeum wind power debates, you will clearly understand the side effects of monocolor.

Given this situation, we are laid back about our nuclear and renewable energy numbers. This is clear. It is difficult for any government to achieve the numerical goals of their preferred energy sources. Even if we do our best with nuclear power and renewable energy under a national agreement, the supply and demand of the energy we need is far from sufficient. It is as if being financially constrained but still deliberating between ordering Tangsuyuk (sweet and sour pork) or Palbochae (stir-fried mixed vegetables), reflecting the dilemma of having limited resources and yet being torn between different options. We are a country that has built the highest-density energy society in human history by relying on foreign resources without any unique domestic resources. We must harness unsolicited energy. However, considering the climate regulations that will be tightened in the future, the flow of de-fossilized fuels must be accepted. This is also an undesirable situation.

Here comes another leadership issue. Currently, national leadership, which changes every five years, is bound to face many difficulties in a situation where long-term public-private partnerships must be guaranteed. However, it is true that when we look at the activities of our elected officials, it is difficult for us to expect political leadership that solve this problem right away. We are clearly a developed country, but we make decisions in the field as if we are a developing country. This phenomenon of hunger is widespread all over the world. At any rate, this gap will prevent us from solving our collective problems. However, we do not need dictatorial and enlightened leadership like in the past; we need new leadership that can stimulate healthy investment and technological innovation. Only then can we remain a developed country.

Therefore, the key to success in green racing is not choosing a good candidate. A good politician is impossible. Thus, the public must take the initiative. Citizens should go ahead and support fuel diversification policies, advocate green racing, and accept fee increases. We must support the companies that have shown results in green racing. Monocolorists must be scolded to prevent them from making false claims. At the very least, citizens must form a consensus and act directly to further enhance Korea's manufacturing capacity. The tyranny of monocolors must be restrained. The monocolors that hinder green racing are clearly our target.

Chapter 3

The Confusion of too Many Old Energy Plans is Also an Enemy

I attended an expert meeting at the International Energy Agency in the early days of my social life. It was only then that we saw with our own eyes the expertise and capabilities of developed countries and reduced our limitations. It was amazing that we surpassed the accumulated capabilities of many developed countries. At the time, I thought that our success lay in our ability to make medium to long-term plans. This is because formulating and executing a five- or ten-year plan in one month is impossible in developed countries. Despite frequent trials and errors due to our laid-back plans, we were able to quickly respond to changes in conditions, make plans, and demonstrate concentration. Of course, the foundation of Korea's dynamism lies in our companies and citizens.

It is true that our national planning capabilities have contributed greatly to our achievements. To be sure, the biggest leadership that made Korea strong over 60 years was national-level policies, and more specifically, the medium to long-term plans of various governments. Although there were liberalization measures such as import/export liberalization measures and the liberalization of the petroleum industry in 1997, it should be assumed that the leadership of the planned economy was still effective. Also, it's safe to say that all these trends were an extension of the 1962 system in a larger framework. However, these existing systems are not efficient given the current situation. The planned economy is only a legacy of developing countries.

However, at a time when climate regulation is operating as a global standard and countries are turning to protectionism, policies are becoming important again. Bill Gates also mentioned in his climate-related book that not only markets and technology but also policies are important. However, since policies are realized through planning, planning is still necessary. Therefore, we must also secure a new planning system suitable for the new era. Of course, the premise of this new approach is the solution to the two enemies I had talked about earlier. In a situation where charges are artificially suppressed and the energy mix is disputed, no planning system will work. Let's start thinking about how to innovate planning capabilities under the premise that these two enemies no longer exist. It was efficient and dedicated bureaucrats who had led this planning. Bureaucrats use powerful organizations and budgets to predict the future and set goals. An action plan to concentrate appropriate national capabilities must be established and executed. This phenomenon is particularly prominent in the field of energy.

Recently, however, it is the commitments of elected officials that have begun to replace the planning function of these bureaucrats. Overly elaborate plans based on expertise are being replaced by vision-based commitments. The energy sector, which has successfully led the strongest planned economy, is the target of the strongest commitment. However, the problem is that the commitment mobilized before the election has no choice but to take a four-year or five-year perspective. However, the vision or planning period for the climate or energy we need is 30 years until 2050. Therefore, it is impossible to fully entrust a country's medium to long-term plan to fulfil a commitment with a lifespan of several years. This is especially true of energy and climate policies. Energy needs a method that presupposes agreement between political parties. Therefore, governance such as the Sustainable Development Committee, which is a discussion structure initiated by President Kim Dae-Jung, is necessary. Of course, the names have continued to change in the Green Committee, the Carbon Neutrality Committee, etc.

In addition to this political point of view, the difficulty of planning income has increased in practice, and this has led to a situation where the utility of the plan is questioned. Related energy technologies are evolving in a variety of ways on a scale that is too small. Accidental plans, such as delays in the construction of power grids, contribute greatly to turning plans into public opinion. As circular coupling is strengthened, it is also quite difficult to predict future demand for each source. Furthermore, as there are too many stakeholders, the ability to execute plans is another challenge. As market rules and the like change dogmatically, barriers to trading are often introduced. In conclusion, it is difficult to make plans, and it is also difficult to execute them. Depletion of location is the biggest factor that threatens the effectiveness of plans. With regard to licensing, it is the most difficult obstacle. A good example is spent nuclear fuel disposal sites.

Another problem is that Korea has so many energy and climate-related plans. However, there is no distinction between presenting a vision, showing direction, and making an action plan. Vision, direction, and action are processed in one package. This phenomenon began when an attempt was made to mechanically match the national reduction target, which was high, with the basic power supply and demand plan, which was a low-level plan, at the time of the second basic energy plan. It was an attempt to respect the legislative intent of the Basic Energy Act regarding the discipline between the basic plan and the sub-action plan. A complete, top-level basic energy plan was created, but this bundled planning system made the mistake of unifying ideals and reality into one, resulting in vision, direction, and execution all going haywire. Furthermore, politics has come to function as a channel for controlling bureaucracy. Only NDC will be adjusted uniformly, even at the power plants. It's such an outrageous and funny situation considering that these plans don't even work on the field. The wonderful planning ability that led to Korea's success was self-indulgent.

[Summary of key energy and climate-related plans]

Schemes	**Contents**
Basic Energy Plan	It is the highest-level plan related to energy and was established considering all the macro-directions of energy measures. It includes energy demand and supply forecasts for the next 20 years, technology development, and workforce development plans.

Schemes	Contents
Basic Electricity Supply and Demand Plan	The basic power supply and demand plan is a plan to forecast electricity demand for the next 15 years and establish plans for power facilities and power supply configurations for stable power supply and demand.
Renewable Energy Basic Plan	The Renewable Energy Basic Plan proposes promotion strategies in terms of renewable energy distribution, market, demand, industry, and infrastructure to support renewable energy's leap forward as the main energy source in the future.
Climate Change Response Basic Plan	The basic climate change response plan is a top-level plan related to climate change response and sets the overall goals and vision of climate change policies, such as establishing national greenhouse gas reduction targets for the next 20 years and presenting the principles and directions of subplans.
2050 Carbon Neutrality Scenario	The 2050 carbon neutrality scenario suggests the direction of sector-specific policies by forecasting Korea's future vision and sector-specific transitions through the realization of carbon neutrality in 2050.

[Economic Information Centre, Economic Policy Time Series Service, 2023.02.11. search]

Source — https://epts.kdi.re.kr/

It is now necessary to loosen the mechanical integration between vision presentation, direction guidance, and action plans. As an example, NDC for climate response is of course appropriate for visionary use. However, it is necessary to have a process that can receive support from the industry rather than being led by an environmental group. Of course, it is appropriate to carry out a basic energy plan when directing the energy policy to implement this. Since the basic energy plan proposes various policy instruments other than NDC, it is desirable to minimize the numerical goal-setting function where possible. It is appropriate to use the Basic Energy Plan as a tool to respect NDC, suggest a medium to long-term direction for implementing it, and build consensus among citizens and stakeholders. This conforms to the nature of the original basic plan. Having participated in formulating the Energy Act at the time, I think that the discipline provisions for subplans under the Energy Act were directional rather than detailed figures. A consistent direction is important. Figures are results realized by operating in the market.

In particular, the name of the Basic Plan for Electricity Supply and Demand, which has the strongest execution, has been changed continuously, but it is 60 years old in terms of content and methodology. Also, it's been about 20 years since they tried to temporarily utilize the power market called CBP. It has been pointed out for a long time that it doesn't fit well under the current circumstances. Thus, many consider the basic electricity supply and demand planning to be useless. There is also a need to transfer permission for the entry of various power generation projects to market rules now and switch to a function that predicts, and forecasts demand as originally planned at the time of the restructuring discussion.

In the new planning system, it is important to present a vision for breakthrough technologies proposed by the LEDS Expert Forum discussed earlier. It's foolish to make a small 15-year plan to build a few power plants now. The most important thing about reorganizing the planning system is to set the direction for how to reorganize energy sources in the future. The focus must be on expressing a position on the implementation period of breakthrough technologies and proactive demand generation for this purpose rather than planning the volume of power plant construction. Also, specific details about incentives should be presented. To further simplify, not only the plans of the Ministry of Trade, Industry and Energy related to energy technology but also the technology-related plans of various government departments such as the Ministry of Science and ICT, require measures to raise their status in carbon neutrality and energy policies. Recently, the pace of innovation in energy technologies that were evenly distributed is rapidly increasing. Green racing is in full swing. Therefore, it is very reasonable to integrate energy policies and industrial policies by making energy technology policies nationally selected policies. Furthermore, efforts to closely link science and technology policies are needed. A new era of industrial policy has arrived around the world.

Chapter 4

There are Groups that Must be Persuaded to Join

There are groups that are usually mistaken for green racing enemies. However, in my experience, they are important partners in implementing green racing.

First, the most important potential Green Racing partner is Korea's consumer groups. Consumer groups with a long history recognize not only the simple protection of consumer rights and interests but also the historical responsibility of being a responsible consumer. Consumer groups do not claim the right to buy cheaply at random. I have a strong sense of buying at my own price. Normally, consumer groups evaluate that they are only interested in lowering utility bills, but it was the very same consumer groups that opposed the one-sided rate reduction plan at a public hearing on the progressive reorganization of electricity bills aimed at lowering charges a long time ago. They are a valued supporter of Green Racing.

There are partners who usually receive the same reviews as Nimby's spokesperson. These are the residents who seek various benefits based on location. If they want subsidies, they are more strongly opposed. Some act as grant hunters with licences as collateral. They are the local public and private sectors operating as a single interest group. Among them, there is a group called the indigenous Toho Force, centred around local builders. When there is a shortage of special industries in a region such as Korea, it is often necessary to rely on subsidies from the Centre. It is necessary to understand the irrationality of the allocation process surrounding these various subsidies. This is because there is an exquisite order everywhere.

It is inappropriate to dismiss this dilemma only as local greed. The locals are just doing their best for their own benefit. They are smarter and more rational than groups steeped in flimsy ideologies. We can have as many conversations as we want. I am confident that residents will accept this and cooperate if rules and regulations are set proactively. The Central government and public enterprises were flimsy. Isn't everyone selfish? Most of these conflicts tend to arise from a lack of proper regulations and procedures. Problems can be solved by setting appropriate goals and presenting rules in a timely manner.

First, there is a tendency to view public groups that settle for the existing system, especially energy companies, as inefficient. In the case of the public sector, unique sincerity also causes the side effects of maintaining the past systems. Conservatism related to public sector purchases acts as a barrier to technological innovation. There were efforts to innovate public procurement during the Moon Jae-in administration, but the results were found to be lacking.

However, in the end, they can be expected to play a huge role as a core force to support the greening of the industry if they have the right motivation. I am proud that the Korean public sector still has a sense of patriotism. Public sector workers can easily intervene in somewhat risky or uncertain projects. This is because there are usually few laws where people are disadvantaged even if they fail. However, the case of private companies is different. Therefore, a somewhat adventurous project that creates a market that does not currently exist but is expected to grow in the future can achieve surprisingly good results when promoted by the public and private sectors through the division of roles; there are many such examples. Public intervention can be of great help to companies if channeled in the right direction.

And finally, there are various economic entities that want to settle for the existing low fees! They are not evil; they are just rational economic agents faithful to their interests. They are mainly traditional industries (in marginal situations) that continuously try to rely on low energy costs and consumers (mainly farmers) who have become accustomed to various subsidies. However, persuading them to innovate on their own is Green's leadership itself. In this respect, it is necessary to make efforts to some extent to compensate them for the disadvantages that will occur during the transition process and give them time to adapt. If possible, it would be nice to give them a chance to evolve on their own.

In the past, new lighting projects and high-efficiency motor projects have also provided additional research funding and technology-sharing opportunities for small fluorescent lamp companies and electric motor companies. The government persuaded them by thoroughly talking about the justification for market changes. This is government leadership. They may resist. However, they are not the enemies. Don't be afraid to talk to them. They too want to survive and develop and evolve. Everyone is smart enough. And they were all surprisingly patriotic.

Part 5: A Reflection on Green Racing and More

Ultimately, the key to making green racing a reality is expanding the mission of energy. The problem is that the energy system is still at a point where it is difficult to undergo significant changes. However, green racing is a national task that must be done, and we cannot succumb to the current difficulties. Thirty years have passed since I entered the energy field in 1992 when the Ministry of Power Resources still existed, and I perceived it to be under Military Support Command. However, as time passed and carbon neutrality was discussed, I felt like I had switched to working in a front-line combat battalion. The role of energy is changing from support to combat. And it's changing from a larger framework to an industry that's leading the green race. If you consider what the minimum urgent effort is to carry out an expanded mission, the following two things come to mind.

The first is to suggest a long-term direction for the energy system. An energy system is like a giant oil tanker. It's like running a small motorboat; it's difficult to reduce it to a short-term policy. However, energy technology is becoming more diverse, and the pace of innovation is becoming faster. Also, geopolitical uncertainty is growing, as seen in the war in Ukraine. In this respect, the most important concern is the direction of the future evolution of Korea's energy system. I would like to suggest four directions from this perspective.

The most important thing to focus on is the rules for investing in and operating a business. This is usually called the market rule. Market rules governing daily transactions are the essence of energy for operators. The current rules of the electricity market are too complicated and uncertain. It's hard for anyone to predict how much profit will be generated. This is true for both public and private businesses alike. Furthermore, various markets such as renewable energy, hydrogen, and emission rights are growing rapidly. However, there are many concerns about integration. Furthermore, more complex rules will be redesigned to solve problems in the electricity market. As a result, uncertainty will be further amplified. New market rules are needed that can protect consumer interests, expand investment, induce technological innovation, and ensure supply and demand stability. This is essential to lead the green race.

Chapter 1

The Energy Policy Must be Improved in Four Basic Directions

Innovating the hardware and software of future energy systems, especially power systems, is a proactive measure towards green racing. This is because, in addition to the evolution of the electric power industry, it is necessary to provide RE100 opportunities for our manufacturing industry, and simultaneously provide track records and stimulate demand for innovative products in individual related industries. There's nothing new under the sky. If we look at our past failures and regrets, it's easy to get the understanding and sympathy of consumers and citizens. Let's attempt to understand our situation and identify what kind of preparedness we should arm ourselves with.

A long time ago, when we were young, we heard nagging comments related to "a country without a drop of oil." Additionally, the phrase "Turn off one light" may have originated from the words of our nation but has gradually become ingrained in our upbringing as part of family education. When we were aiming for industrial sophistication, we faced many challenges because we were a resource-poor country. The oil crisis that occurred concurrently may have further strengthened our desperate feelings about this energy. Amidst serious debates between government departments, someone asked, “Aren’t you trying to save energy?” The debate often came to an end with just one word. As such, the national consensus on energy-saving was clear. However, this healthy obsessive sense of saving energy quickly disappeared. Electricity has become a convenient energy source. It’s hard for the old enlightenment slogans to work. Presently, a more sophisticated method, such as culture and art, is needed to revise our standards of consciousness. Culture and art have such great power. Because Korea is a developed country now.

Currently, electric energy accounts for about 20% of the total in Korea, and the rest is still mainly based on fossil fuels such as petroleum. However, regarding climate, for decarbonization, the proportion of electrification will inevitably increase. But we are already at the highest density, and the expansion of electric energy is quite difficult in many ways. Even now, power plants built on the East Coast are unable to operate due to a lack of transmission networks, whether nuclear or coal. The west coast has a sky filled with transmission lines, and further construction may be challenging. The area is already congested. After the Miryang crisis, the construction of additional social infrastructure such as the radio station in Korea was limited. Now, in addition to solar power and offshore wind power, almost all social infrastructure facilities, such as small-scale power plants, substation facilities, and waste treatment plants, are blocked in most regions. The available sites for such infrastructure are practically exhausted.

In fact, this situation was predicted a long time ago, and in the mid-1990s, the Ministry of Energy sought to shift the focus from supply to demand management (Demand Side Management, DSM). There may have been a belief that it would be difficult to build a continuous supply facility. They had already anticipated the depletion of location. Of course, energy efficiency projects continued functioning through the Korea Energy Corporation established in 1980, but it was a strong reversal attempt to turn the energy supplier itself into a demand management agency. Therefore, they switched to a policy of demand management and implemented legal regulations so that KEPCO, which had a monopoly on power generation, transmission and distribution at the time, also invested in reducing electricity demand. The government took the initiative and decided to drastically reduce the number of power plants by millions. Officials at the time tried something surprising even for current environmental groups. However, after considerable backlash against reducing sales to electricity retailers, this policy degenerated into a very small-scale subsidy project. Even now, the system is respected and is reflected in the basic plan for electricity supply and demand, but its effects are minimal. Demand continues to increase, and positions are becoming more and more exhausted.

Recently, however, the IPCC announced that decarbonization of the supply chain alone is not enough, and demand must now be reduced in earnest to satisfy the 1.5-degree scenario. It is not about reducing energy use while maintaining demand through existing efficiency improvements. It is a shocking suggestion that global growth should be reduced by reducing demand, that is, the market. It's a warning that climate change is so serious.

However, whether it's energy savings or high efficiency, Korea still has a lot of potential. If you go to the city, you can see several stores that have the air conditioning and heating on and open their doors at the same time to attract customers. There is still a lot of demand for energy-saving facilities. On the fuel side, electrification is inevitable, but efforts to reduce the burden on facilities by maximizing the efficiency of electrification are needed first. It is normal if demand management comes first, and fuel transition comes after that. This is the 'priority principle of demand management'.

The second is the issue of decentralization. A memorable scene is the content of the Third Basic Energy Plan established during the Park Geun-hye government. At the time, it was difficult for environmental groups, nuclear energy groups, and public-private committees composed of experts to reach an agreement on the weight of nuclear power plants. The agreement was not respected by the Moon Jae-in government, but in fact, there was nothing to say. At that time, there was an additional agreement on decentralization and tax system reorganization in the nuclear power plant share agreement. Some tax reform was realized, but the decentralization agreement was immediately abandoned by the government. Therefore, there is a limit to making claims to abide by the nuclear power plant agreement made by the government.

Also, with regard to the CFI project implemented in Jeju Island, KEPCO's transmission executives and a group of experts voiced their concerns about the anticipated network saturation. However, the meeting came to an end after they were ignored by government officials of the Metropolitan Government

Office. If they had accepted the experts' suggestions at the time and responded, the current renewable energy output limit (Curtailment) turmoil would not have occurred. Officials should always respect the opinions of bitter experts. I've heard stories of Japanese government officials lamenting during the Fukushima Bay nuclear disaster in Japan. There are two types of experts in Korea – misuse and dance. There have been concerns about network issues for a long time. Networks are such an important and serious issue.

At the time the basic plan for secondary energy was established, the basis for the strong arguments of experts over decentralization was concerns about the saturation of transmission networks across the country rather than Jeju. According to the investigation at the time of the Miryang crisis, it was confirmed that residents living along the transmission line suffered actual property damage. For example, compensation is given to residents living near a power plant. There are exceptions, but usually, there's an increase in the price of surrounding land. This is because regional development, such as road construction and lodging facilities, is also required. However, while transmission damage to transmission lines is exaggerated, the damage caused by hitting a trade cliff is clear. In terms of electrical engineering, the point that there was a limit to electric transport was rather convincing. As a result, costs were higher, small-scale regional conflicts were further expanded, and difficulties in network operation were anticipated, but decentralization was an inevitable choice. If the government had faithfully implemented the decentralization agreement at the time, many complicated and painful consequences we are facing today would have been reduced. It's a pity that the decentralization agreement could not be realized.

The third story is about sector coupling which is recently being discussed in earnest. Our challenge is to further promote electrification to achieve carbon neutrality, expand demand for electric vehicles, DR, etc., and expand the share of nuclear power and renewable energy on the supply side. To this end, power grids must be innovated to be smarter. At the same time, it is necessary to expand the demand for hydrogen and increase collaboration with thermal energy, etc. However, our country is already cramped. It's full of power grids, gas grids, and heat pipelines, and now this infrastructure needs to be further expanded, and a new hydrogen network needs to be built. This is the real challenge.

Considering the country's limited space for power, gas, heat, and even hydrogen in the future, additional infrastructure expansion is expected to be quite difficult. Therefore, it is necessary to make efforts to maximize efficiency by utilizing limited individual infrastructure optimally. Efforts to maximize the efficient use of energy through mutual flexibility can be described as sector coupling. As a result, all combinations of trading methods, returns, and innovative technology will change. In this reality, we cannot ignore the influence of central departments and leaders of basic organizations. Therefore, rules and road maps that can receive support from consumers, businesses, and investors are even more urgent.

It is necessary to overcome this lack of space by operating infrastructure in a more integrated manner, expanding mutual flexibility, and allowing transactions while developing completely new services in the process to provide newer, more value-added services. Also, it must be possible to use the existing gas station space as a complex charging space that can be used by electric vehicles and hydrogen vehicles, like an energy superstation. At the same time, licensing systems that are regulated on a per-source basis must also be integrated to enable such integration. Of course, X2X technology such as V2G, which crosses related circuitry, must be secured early. This is also known as sector coupling. Countries that achieve legal, technical, and positional integration first can win the green race.

Pink hydrogen using nuclear energy, green hydrogen using offshore wind power, and gas-based blue hydrogen using CCUS are sector couplers themselves for such integration. A newly realized hydrogen society will be a key new investment opportunity for building a more integrated energy system that takes these past events into consideration.

The last thing I want to emphasize is the issue of creating demand for innovation, which is the government's responsibility to support green racing. On a global level, green racing is essentially a battle over which country will be the first to become a key technology supplier. Korea can become a newly emerging arms supplier to the West with K Defence. Should we take the lead in supplying innovative technologies such as electric vehicles, ESS, CCUS, and hydrogen reduction steelmaking? For our manufacturing capabilities to win here, the government's role is to generate demand for innovation to secure technological advantage and mass production competitiveness within Korea at an early stage. Institutionally, this means an accelerated subsidy system, mandatory technology regulation, or public purchasing.

There is no separate consideration for Korean companies in Korea's energy market system. However, they can enjoy some preferential treatment in the detailed calculation of subsidies. It is true that our manufacturing capacity is insufficient to secure track records to preempt the global market or secure initial innovation demand at the level of welcome goods. Even if it costs a little more money, it is preferable to provide Korean companies with export opportunities using domestic technology rather than importing foreign products. In the past, large-scale subsidies were provided for secondary batteries. At the time, I had objected to this, but if you look at it now, it helped secure our current global position as regards secondary batteries. Wisdom to make good use of subsidies is necessary, and it is preferable to incorporate it into a market system where possible.

Let's talk about three grant scenarios related to this. If there is no government distribution support, there are three types: Subsidies for the cheapest technology and additional subsidies for domestic products. Let's say these are the 0, 150, and 170 won scenarios, respectively. The 170 won scenario is a bit expensive, but it provides an additional subsidy of 20 won for domestic products. Of course, I'm not trying to provide additional support for domestic products that have no potential. The 0-won scenario also involves a strategy of waiting for the technology to become cheaper in the future. Of these, I think the least expensive option is the one worth 150 won. If you want to distribute imported technology without any industrial policy effect, I think it's better to wait a little longer until prices are meaningfully lowered to distribute it. On the other hand, if some competitiveness is expected, it is wise to try to accelerate domestic technology through effective methodologies even if it costs more subsidies. This is because if global competitiveness is secured through this, the item will become K-Bang's brother. Of course, it takes courage to make such bold decisions and wait. This is the new industrial policy that has been recently revived in a new era by most countries. Look at the US IRA.

The basic direction of innovation in future energy systems should still be demand management principles, decentralization principles, integration principles (sector coupling), and innovative demand generation. This will contribute to carbon neutrality and make it possible to upgrade our power systems and make related industries greener. The remaining homework is to reorganize existing systems, especially planning and market systems so that investments are made in this direction. Of course, it's not easy. However, it's an era's homework that must be solved, and it's the path to green racing.

Chapter 2

Integrates Complex Energy Market Rules in an Easy-To-Understand Manner

Energy is the most representative ubiquitous system that is incredibly large, complex, and operates 24 hours a day. Every day, consumers buy and use energy such as electricity, heat, and petroleum from the market. Complex rules are at play that determine who will set the price of these energies, who will be responsible for the transaction, and how much tax will be paid. In addition, there are myriad regulations and procedures that determine who will import resources from abroad, who will license the business, and what the yield or size of the subsidy will be for the project. The sum of these rules constitutes the energy market and the reality of the energy policy.

Petroleum is a product selected by consumers through competition among gas stations, but in the case of electric energy, it is still a rigid market where KEPCO sells exclusively. The way electricity is traded is determined by the market on the outside, but internally, it is still controlled by the government. The electric power market itself is complicated, but complicated transactions linked to thermal energy and renewable energy are being carried out every day. With the expansion of nuclear power or renewable energy, which is difficult to control, consumers are anxiously and constantly managing tasks that are hard to understand, such as frequency management to prevent power outages, in a highly time-sensitive manner. Considering that there are a growing number of large and small businesses, modifying their interests is challenging. However, this system has become so complicated that the government and experts sometimes get lost.

Currently, the trading of electric energy follows the Cost Based Pool (CBP) market rules. It was a system that was decided to be used temporarily during the restructuring process 20 years ago. Power generation companies sell electricity to KEPCO, an exclusive sales company, and KEPCO sells directly to consumers as a retailer. Currently, Busan, Seoul, and Gwangju all receive the same retail rates in Korea. Although the price of gas purchased from overseas changes rapidly, our consumer prices are very inflexible. At the end of last year, the government reorganized the fee system to link fuel costs, but it just didn't work due to political reasons. The electricity bills in Europe skyrocketed nearly 10 times due to the recent war in Ukraine, but Korea has maintained stable prices and KEPCO is enduring a deficit of close to 30 trillion won. Of course, there are pros and cons to this situation. Electricity consumers have stable businesses, but KEPCO must continue to deal with the volatility of deficit surpluses.

<Korea's Electricity Market>

Korea has traditionally operated a CBP market. CBP is a wholesale power market introduced when the electricity market was established in 2001 and includes the energy market and capacity market. CBP is a unique type of market where bid prices are determined based on cost. Since Korea Electric Power Corporation has supplied electricity mainly from nuclear power and coal, CBP has operated stably in Korea. Recently, however, new markets have emerged as private power companies such as GS, EPS, POSCO Energy, and SK E&S have entered the market and the need to distribute renewable energy has increased. Examples include the Renewable Energy Portfolio Standard (FIT) related to renewable energy, the Renewable Energy Portfolio Standard (RPS), the Hydrogen Energy Portfolio Standard (HPS) system, and the greenhouse gas emission trading scheme (ETS).

FIT is a system where the government purchases electricity produced from renewable energy at a fixed price to support the spread of renewable energy. FIT had a disadvantage in that when the renewable energy market expanded, the burden of the difference to be covered by the government increased, and RPS was implemented as an alternative to this. RPS imposes an obligation on power generation companies with power generation facilities of 500 MW or more to supply at least a certain percentage of the power generation with renewable energy. After 2023, the mandatory supply ratio was 10% of the total. In addition to directly introducing power generation facilities, it is possible to achieve the allocated supply by purchasing certificates from other renewable energy power generation companies.

Recently, to expand the hydrogen economy, RPS introduced HPS, which manages fuel cells separately. The proportion of hydrogen in RPS is 13%, which is not significant compared to technologies that are already highly mature, such as solar power and wind power. Therefore, the goal is to accelerate distribution by independently managing the mandatory supply ratio for hydrogen power generation.

ETS is one of the Kyoto mechanisms stipulated in the Kyoto Protocol. It is a system where the government allocates greenhouse gas emission rights to workplaces that emit greenhouse gases and induces them to act within that range. Workplaces can evaluate greenhouse gas emissions and obtain sales revenue or achieve quotas through transactions between workplaces for extra or insufficient emission rights.

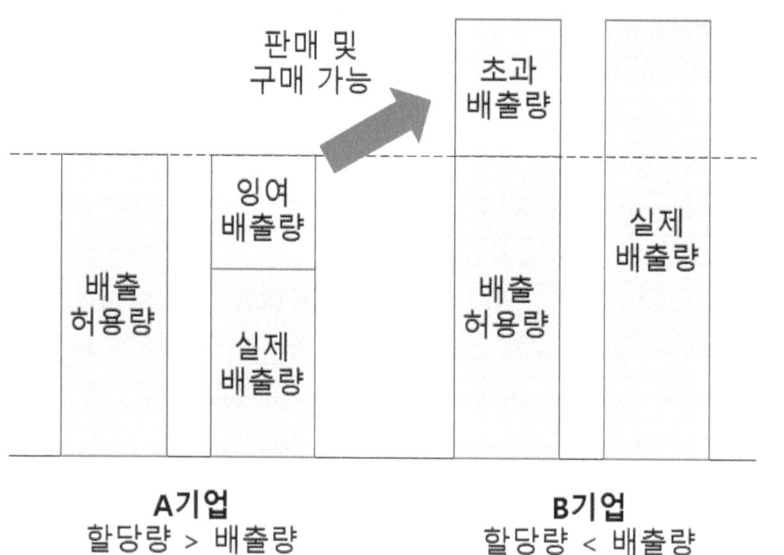

[Electric Journal, Current Market Status and Improvement Direction Reflecting Variable Costs, 2019.11.] [Enlighton, What is the RPS System? 2020.01.] [Korea Economy, Mandatory Hydrogen Generation System] [Korea Environment Corporation, Greenhouse Gas Emissions Trading System]
http://www.keaj.kr/news/articleView.html?idxno=2972 https://www.enlighten.kr/insight/glossary/1044 https://dic.hankyung.com/economy/view/?seq=14705 https://www.keco.or.kr/kr/business/climate/contentsid/1520/index.do

However, the rules of this power market are causing more and more problems. In other words, from the perspective of a business operator, it is difficult for anyone to predict their own rate of return. Profit cannot be determined by one's own efforts. Recently, the SMP upper limit led to a situation where the government directly controls the profitability of operators. Who would want to invest in power generation projects and infrastructure? On the other hand, there is a belief that consumers will spend rationally if retail prices fluctuate all the time. However, there is a possibility that the current low rates will encourage reasonable consumption. That's because it's so cheap. In a situation where investment in climate response must be expanded, the trend will go the opposite way.

This power market is complicated and has many problems. Experts are concerned about attempts to solve problems by making current power market rules more advanced, that is, injecting more complex rules. If there are no consumer price liberalization measures, such as in the oil market, making more rules will not solve the power system's equation.

However, what we must show interest in are other markets that are currently small but are evolving steadily. In addition to traditional thermal markets such as district heating, various local markets are evolving with their own logic, such as the RPS market for renewable energy, the HPS market for hydrogen expansion, the ETS that allow mutual trading of emission rights, and EERS, which has limited specifications related to future efficiency. However, all these markets are siblings of the CBP market, in that they occur in the electric energy production process. Electric Power Market Brothers. If these markets continue to grow independently as they are now, there is a possibility that our energy market will fall into a real ladder-climbing game, where competition for market share becomes the primary focus. There is a great possibility that these brothers, who have become extremely complicated, will confuse consumers who pay money and businesses that invest. Not only the government but even the gamblers in the financial sector will find themselves in dire straits. Even if I invest, I don't know how much I'll earn, so I wonder if the investment will happen.

Electric Power Market Brothers play an important role. These brothers will determine the size of the portfolios, returns, and subsidies for each source of coal, nuclear power plants, renewable energy, hydrogen, demand management, etc. Appropriate mixes are needed to respond to climate change and fine dust in the future, and it is necessary to achieve this by strengthening the integration of these brothers. The adjustment of the mix must be a transaction method that can be made automatically according to market rules. It is these market systems that complete their tasks with money paid by consumers anyway, and it is these market systems that determine the principles of allocation by converting them backwards. The market system itself is also part of the policy. In the end, it is preferable to implement integrated market rules rather than plan the future mix adjustments required for climate response.

What is important in market design is to have confidence in how the market will evolve. It is good to have a road map for the evolution of a strong market system. Recently, the market has become mistrustful of government policies due to frequent policy changes. Energy is the most representative policy issue requiring a long-term perspective. But if policies change every time the government changes, when the officials in charge change, who will invest? To respond to these problems, the government needs to present a road map. I remember that even in those complicated discussions about the restructuring of the electric power industry, only a one-page roadmap remained. There is an urgent need for a roadmap for the direction of the evolution of the energy system that citizens and major stakeholders can realistically empathize with and acknowledge in a larger framework. The greater the consensus, the greater the stability of the market system. However, considering the complexity of the matter, it would be better to slowly proceed with the realization of the market's roadmap over a ten-year period.

Finally, there's one important principle. It's called governance. When setting and changing the rules of the market, the participation of stakeholders who are paying all these costs is important. They include businesses and consumers. Most electricity companies in Korea started their business by obtaining permission under the government's plan, expecting a low but stable profit rate. However, if the government randomly changes market rules from time to time, operators are bound to be confused. Even now, there is tension in the atmosphere that will spark large-scale lawsuits. Furthermore, since KEPCO is listed on the New York Stock Exchange in the US, stories of international lawsuits are often discussed. This level of institutional instability is too embarrassing considering we are an advanced country. At the same time, there is growing concern that this will have a negative impact on large-scale investments needed in the future.

Also, the opinions of consumers, housewives, and factory owners must be sincerely considered. If it is necessary to listen to the opinions of these stakeholders when revising policies or market rules, no one will be tempted to change market rules frequently to carelessly respond to short-term issues. With the stability of these rules, investment will also be carried out more smoothly. The government can also be free from the temptation to intervene in the market as a temporary measure to respond to current issues. Consumers who are the true owners of the market can also protect their rights. Smart, responsible consumers can protect their own conscience and descendants.

Is democracy separate from humanism? I sincerely hope that the day will come when these brothers in the market system will harmonize with market participants, compete in their own way, and contribute to Korea's ideal mix and green racing.

Chapter 3

Green Racing Also Requires Strategic Thinking

Green racing involves competition between countries. Therefore, what to do is important, but the efficiency of achieving it and the strategies that guarantee it are important. In this regard, I would like to suggest three strategic ideas. First, there is the issue of the principle of transformation – how to sort out the problems between existing and new industries due to greening. We need to consider whether the commonly talked about 'just transition' is valid. And secondly, who will take the lead? There is a need for collaboration between the private and public sectors. Finally, there's the issue of the speed of conversion. After all, speed was also a factor impacting the costs or subsidies that countries initially paid in advance. In other words, we are competing with the speed and strength of greening in other countries.

The first question is what to do with the existing energy consumption industry. I have heard an official from the Ministry of Environment suggest that some energy-intensive manufacturing industries such as steel and petrochemicals could be abandoned. Will this approach help national interests and the survival of humanity? The damage to national interests is clear, and globally, the factories that emit greenhouse gases are simply moving to other countries that are cheaper and emit more. Here, I want to focus on a problem that dominates our thinking. It is 'just transition.'

The phrase that encapsulates the anticipation and preparedness for this conflict between the old and the new is 'just transition.' Since carbon neutrality is definitive and irreversible, it is necessary to focus on minimizing damage to industries, workers, and residents that are unfairly harmed during the transition process. However, in the current situation, the term 'just transition' seems too casual. Countries stopped investing in traditional energy and prepared for the future in order to achieve carbon neutrality, but they experienced the strong influence of traditional energy through the current crisis in Ukraine. Concerns about whether 'just transition,' which presupposes the preemptive exit of industries based on the existing energy system, can guarantee its effectiveness is deepening.

This is an even more serious issue, especially in mid and post-industrial manufacturing countries such as Korea. The manner of response after a preemptive exit can indeed lead to losses in both the present and the future if not carefully managed. We must prepare for the future with funds secured by the current system. For example, the reality is that it is necessary to invest in the future with special energy accounts and electric power industry infrastructure funds currently provided by the energy industry. Therefore, a harmonious transition is needed rather than a discontinuous transition.

The key to a harmonious transition is to maintain the current system and invest in the future with funds derived from it. It promotes innovation in traditional manufacturing industries such as shipbuilding, steel, and petrochemicals, led by a low-carbon energy system based on nuclear power plants and renewable energy, and digitalized electric vehicles and hydrogen trucks. Simply put, in a tight household, it is necessary to build a home with two houses of tradition and innovation for a considerable period. We need more money. The only solution to overcome this is a bold innovation investment led by the private sector to promote innovation in the energy system and manufacturing industry.

So, who will focus on the second issue? Currently, it is private. When considering the scale of climate response technology, large companies must play a realistic role. However, there is a need for large corporations to be reinstated automatically in South Korea. And once again, the long-standing debate over the spillover effects will be sparked. In this respect, it is necessary to consider a system that matches the interests of large companies with national interests. With the cruel reality of the international community revealed due to the war in Ukraine, can the 'good ESG' continue to survive? ESG is a very effective and transparent basis for them to fully express their strength domestically and globally, so it must continue. In conclusion, I think this morality will play its role even in the dizzying Sengoku era. This is because only true ESG (or CSR, whatever the name) can support the expansion of Korean companies. Also, even under domestic egoism and protectionist trade, morals (such as ESG) are abstract but will ultimately contribute to securing competitiveness.

However, for the private sector to efficiently compete with companies in competing countries, the role of the public is important. The public must play an active role in proactively resolving obstacles such as licensing, safety, and residents' negotiations before the private sector invests. This combat engineering role of the public will be necessary in public-private collaboration, and the Korean public has the potential to handle such things. Also, the public sector must generate innovative demand. Through this demand, our companies can focus more stably on technological innovation. Investments are not possible with the current licensing and yield determination methods. The public must take on this major role. In this way, it must create conditions for our companies to innovate. Also, consumers must be prepared to pay a certain amount of money. Governments and the public in every country are doing these things. Diplomats from developed countries are also doing these things. However, the Korean public must adapt to these changes faster than the public in other competing countries.

To do this, it is necessary to arm the public sector with a high degree of transparency and expertise. This itself is a process of past reform. Fortunately, these difficulties are the same in other countries. If you have the courage to walk a little faster than them, you can achieve unexpected results. Fortunately, the current government is active on this issue. I emphasize once again, to achieve carbon neutrality, Korea must pursue a policy based on strict national interest that attempts to transition while maintaining the status of the manufacturing industry, the foundation of our survival. Only such a strategy can enable Korea to effectively contribute to carbon neutrality, which is a global issue, and respond practically to the current economic crisis.

Finally, the speed of conversion is critical. We tend to view existing and future systems in a hostile way. However, we have the experience of building by fighting. South Korea has advanced at a tremendous pace. The speed of our transformation is highlighted by the Korean symbols ‘Dynamic Korea’ and ‘fast fast’. We are fast. The speed of green racing is also a factor in Korea’s unique global competitiveness. The introduction of gas promoted during the Jeon Doo Hwan government is a typical example. The country has shown its ability to quickly handle processes such as policy decisions on gas introduction, gas introduction contracts, infrastructure construction such as gas pipes, etc., establishment of administrative systems such as gas companies, and large-scale gas demand generation. Gasification was literally promoted at the speed of electro-petrification. I hope that this ability will be demonstrated in building a hydrogen society in the future. The key to this speed challenge is, of course, a bold investment. If you look at it strategically, a bold investment in various infrastructures is critical.

Underlying this bold investment is a firm decision on fuel diversification policies. There is a political and public effort to increase energy security through fuel diversification in the medium to long term and to satisfy consumers’ desire for advanced energy even if there is some risk of redundant investment in the short term. A strong fuel diversification policy is the core of Korea’s energy policy. Also, a strong diversification policy guarantees investment, and this belief once again guarantees the speed of transformation. This is an age where courageous leadership that pushes for such diversification is essential. The core of a harmonious transition is a strong diversification trend and the rapid challenges associated with it. You must overwhelm your rivals with a quick transition like Blitzkrieg. Through this, we will become a country that successfully achieves carbon neutrality and manufacturing innovation together.

As a final note, I would like to emphasize that the key to success in green racing ultimately lies in national participation. In this respect, achievement is currently being neglected, but there is one thing we are most proud of when it comes to green racing. It is the forest greening project. If you watch movies from the 1960s, you can often see Mindong Mountain in the background. All the trees on the mountain were cut down because they needed firewood. However, we planted one tree at a time to create the green forests and mountains we have today. I literally planted them one by one. When I was in elementary school, everyone went out to the mountains on Arbor Day and not only planted trees but also caught insects intermittently. And at a very high speed, we made the mountains all over the country green. This is one green racing achievement we can be proud of.

Finishing Here

We've looked at our achievements and crises so far. We have managed the most brilliant achievement since Grandpa Dangun founded the country. We have also looked at the crises that will come our way, and climate change, which is the most serious issue of them all. We have looked at ways to maintain our status as a manufacturing powerhouse. I have defined the three internal enemies that are most dangerous to maintaining our wealth. We must maintain and protect our daily lives and wealth in changing conditions. To this end, let's summarize one more important issue.

We need to start 30 years of long-distance racing to reach the 2050 carbon-neutral goal. Greening is also a sport we have been good at. We will continue to participate in this race. If we enjoy a stronger, more stable, and more humane Korea, and if we succeed in innovation and win the green race, we will unequivocally become an advanced country that faithfully contributes to the common purpose of humanity. Since our companies are competing in the global market, they are already willing to participate. However, experience in the domestic market, track records, and demand on a certain scale are essential conditions for our companies to compete in the global market. Therefore, it is essential for the government to make good use of the domestic market to enhance the competitiveness of our companies. We also need strong business-friendly measures like Europe implemented in response to the US IRA Act. Korea's current climate-related domestic market policies are cluttered. This will only cause confusion for our companies and foreign investors. This is the crisis that is currently ongoing.

Of course, in addition to the climate issue, we will face many difficulties and trials in the future, such as ageing, population cliffs, pension depletion, political conflicts, and the military crisis between North and South Korea. As mentioned above, we have overcome several challenges and achieved great things. The crisis we are facing now is of a different kind. Amidst these countless conflicts, we must reach a clear consensus despite the conservative side of progress. The consensus is that we can protect and grow our wealth only by maintaining and innovating the manufacturing capabilities we have developed like Altoran. At the same time, it is even more urgent to secure energy resources to feed manufacturing capacity that still needs to consume huge amounts of energy. Efforts to secure independent resources such as nuclear energy, demand management, and renewable energy are essential, and innovative resources such as hydrogen energy will be suitable transformation materials for us. The security of energy resources, which are likely to become more vulnerable due to future climate issues, etc. must be protected with special vigilance. These two virtues are irreplaceable. Only then will we develop further. We must continue to develop these two core assets. Only then can we survive. This was the premise of our country's miraculous development and is the most valuable asset that will sustain our wealth.

However, to win this war, several important agreements are required. One is that green racing requires huge investment at the national level, and to implement it most efficiently, we must be prepared to normalize energy charges. Market-based energy rate normalization works temporarily as a result of a decline in the competitiveness of traditional manufacturing industries, pressure on inflation, and a burden on people's daily lives. However, the normalization of energy charges will eventually contribute to fostering industry, creating jobs, and ultimately reducing energy security and burdens. From these two balances, we must now find a new balance suitable for the climate era. There is a great possibility that preferential treatment will be given to some companies, especially large companies. We must make one more choice. We must maintain our society with organic energy, even if it is somewhat expensive. Consumer consent is required for this.

Another key point is to be free from disputes over the energy mix. Excessive bias, such as the demineralization of the Moon Jae-in government in the past, is contrary to the Republic of Korea's energy diversification policy. Meanwhile, the ongoing debate on renewable energy is also causing serious problems. Sustainable energy diversification policies that are eco-friendly must be maintained for energy security and climate response. Towards this end, the political support of the people for diversification policies must be firm. If these issues are resolved, the flexibility of energy and industrial policies will be guaranteed, and Korea will once again be able to further enhance its status as a manufacturing powerhouse.

We must reorganize our energy system and at the same time arm our manufacturing industry by rearming our energy policy with a new industrial policy. To do this, we must innovate our energy policies and energy systems. We need to change faster and more innovatively than other countries. Korea's many energy planning and market systems must be innovated. Through this, it is necessary to provide an atmosphere where businesses can invest to their heart's content. Currently, we are very concerned about our response. I feel apprehensive as our domestic political conflicts get entangled, reminiscent of the eve of the Imjin War. Aren't the activities of the amazing global tech companies in Silicon Valley such as Apple and Amazon immediately shrinking due to the complexity of energy resources caused by the war in Ukraine? This trend will be further strengthened when climate response is in full swing. The world has now entered an era where energy policies are industrial policies. This is the essence of green racing. In short, we are regressing.

Korea's energy system has successfully carried out the traditional task of stably supplying energy to factories and citizens. However, in the climate era, the mission of energy has expanded even further. The energy system's extended mission is to first replace the overall physical infrastructure and energy mix of Korea's energy with a more efficient, low-carbon one. At the same time, it is necessary to promote the greening of manufacturing industries that provide relevant technology. We must support the speedy transformation of all these processes. In a nutshell, we must take responsibility for Korea's energy supply and demand, supporting and leading the green race in our manufacturing industry. This is the new extended mission of our energy system going forward.

Overcoming the three enemies I had talked about earlier is the premise of everything. Absolute support from consumers and citizens is necessary to protect Korea's manufacturing industry from the impact of climate change and move forward as a developed country that is stronger through greening and innovation in the manufacturing industry. Consumers must make it clear that they will pay a fair price for energy without being shaken by political populism. The public must also make it clear that they will not vote for energy monocolorists. The role of expert groups is also particularly important when it comes to government planning. Politically, we must distance ourselves from partisanship. Consumers, citizens, and professionals must protect and support Green Racing with more dedication. The rest is up to the public and private sectors.

If we maximize innovation and investment in green racing, surely we will become a more advanced country. Furthermore, putting national interests first, which aims to maximize Korea's interests through such innovation, will eventually contribute greatly to climate change issues, which is a universal challenge. Maximizing national interests through innovative investment is a path for both humanity and polar bears. We can truly become a developed country if we follow this path.

www.ingramcontent.com/pod-product-compliance
Ingram Content Group UK Ltd.
Pitfield, Milton Keynes, MK11 3LW, UK
UKHW041638190726
13854UKWH00006B/2567

9 798890 268594